Piezoelectric Technology

This book explains the state-of-the-art green piezoelectric energy harvesting (PEH) technology. It highlights different aspects of PEH, starting right from the materials, their synthesis, and characterization techniques to applications. Various types of materials, including ceramics, polymers, composites, and bio-inspired compounds in nano, micro, and meso scale and their recent advancements are captured in detail with special focus on lead-free systems. Different challenges and issues faced while designing a PEH are also included.

Features:

- Guides on how to harvest piezoelectric energy in a sustainable manner.
- Describes related figures of merit for piezoelectric energy harvesting.
- Covers synthesis of piezoelectric materials in the form of bulk, single crystal, nano, and thin/thick film.
- Includes pertinent advanced characterization techniques.
- Reviews piezo-energy harvesting devices and structures.

This book is aimed at researchers, professionals, and graduate students in electrical engineering, materials, and energy.

Piezoelectric Technology

Materials and Applications for Green Energy Harvesting

Swetapadma Praharaj and Dibyaranjan Rout

CRC Press
Taylor & Francis Group
Boca Raton London New York

CRC Press is an imprint of the
Taylor & Francis Group, an **informa** business

Designed cover image: © Shutterstock

First edition published 2024
by CRC Press
2385 NW Executive Center Drive, Suite 320, Boca Raton FL 33431

and by CRC Press
4 Park Square, Milton Park, Abingdon, Oxon, OX14 4RN

CRC Press is an imprint of Taylor & Francis Group, LLC

ISBN: 9781032329062 (hbk)
ISBN: 9781032329079 (pbk)
ISBN: 9781003317289 (ebk)

DOI: 10.1201/9781003317289

Typeset in Times
by Newgen Publishing UK

Contents

Preface

The worldwide energy crisis and environment pollution caused by soaring levels of non-renewable energy consumption has fueled the need for alternative energy technologies in the present day. These include harvesting parasitic energy from ambient surroundings such as heat, fluid flow, electromagnetic radiations, mechanical vibrations, and in vivo energies, including heartbeat, limb motion, body temperature and so forth. Scavenging such energies opens up avenues for driving low-powered electronics, especially wearable and implantable devices, smart phones, wireless sensor networks and so forth. With this motivation, the domain of energy harvesting has been captivating both industrialists and academicians. Researchers from all over the globe are focusing on developing ways to exploit the potential of these witty technologies in order to create a smarter world. Among all, piezoelectric energy harvesting is very effective in encapsulating ambient mechanical and vibration energy and transforming it into electrical power, as it is solely based on intrinsic polarization without the requirement of any separate voltage source. Moreover, piezoelectric transducers are quite durable, reliable, and extraordinarily sensitive to external stimulus and possess faster response times and higher energy density. Additionally, the output power scales with $V^{4/3}$ (V refers to volume) in piezoelectric conversion technology. This inherent capability allows simpler and mini-architectures preferable in MEMS scale devices. The performance of these harvesters can be evaluated on the basis of various figures of merit, such as piezoelectric charge constant "d," voltage constant "g," electromechanical coupling factor "Q_m," permittivity "ε," power output and strain, and so forth. These figures of merit are highly controlled by the genre, size and morphology of piezoelectric materials. In this book, attention is focused on four categories of materials at different length scales (Ceramics, Polymers, Composites and bio-inspired compounds) and their characterization techniques. Further, we shall also make an effort in this book to record a wide diversity of novel piezo-materials that in recent years have stepped into the arena of energy harvesting . It may guide the researchers to navigate the utility of this category of materials from laboratory scale to application at an industrial scale, which is still a long way to walk through.

In the later part of the book, we review the recent advancements in the field of piezo/vibration energy harvesting. This will be demonstrated in different scales ranging from several m^2 piezoelectric floors to submicron nano arrays. Finally, we will discuss various challenges that need to be addressed for achieving clean/green self-powdered systems with implementation of piezoelectric energy harvesters.

The authors take this opportunity to extend their heartfelt gratitude to Kalinga Institute of Industrial Technology (KIIT) Deemed to be University, Bhubaneswar, India, and the beautiful people associated with it.

About the Authors

Swetapadma Praharaj is currently working as Associate Professor in the Department of Physics, School of Applied Sciences, Kalinga Institute of Industrial Technology (KIIT), Deemed to be University, Bhubaneswar, Odisha, India. She obtained her M.Tech in Metallurgical and Materials Engineering from the National Institute of Technology (NIT), Roukela, and was awarded with the Institute's silver medal for her performance. Thereafter, she completed her PhD in Physics at KIIT, during which she honed her expertise on the development of lead-free piezoceramics and improvement of their figures of merit. To her credit, she has published in 30 peer-reviewed journals.

Dr. Praharaj applies her expertise to best tailor the functionalities of eco-friendly ceramic materials that are the materials of choice in electronic components and high-tech machines. She has a keen interest in developing high-strain and temperature-stable high Tc ceramics (applicable in actuators, sensors, and capacitors) by compositional engineering. She is proficient in evaluating the performance-based figures of merit of materials in terms of S-E, P-E, Dielectric, and impedance spectroscopy. Currently, Dr. Praharaj is working on designing energy storage and harvesting devices using lead-free piezoelectric and supercapacitor (electrode) materials.

Dibyaranjan Rout is currently working as Associate Professor at the School of Applied Sciences (Physics), Kalinga Institute of Industrial Technology (KIIT), Deemed to be University. He obtained his PhD in Physics from the Indian Institute of Technology (IIT), Madras, and subsequently took a postdoctoral position in the Department of Materials Science and Engineering, Korea Advanced Institute of Science and Technology, Korea. He also served as a Research Assistant Professor in the same institute for two years. To his credit, he has published more than fifty papers in peer-reviewed journals. Under his supervision, five PhDs and ten master's theses have been awarded.

Dr. Rout's research interest encompasses the structure process–property relationships exhibited by functional/smart materials, particularly lead-free piezoelectric/multiferroic/high-Tc relaxor materials. He is keen to design and develop such novel and eco-friendly materials in the form of bulk ceramics, nanoparticles, nanostructures, thick films, fibers, and nanocomposites using ease and cost-effective processes (solid state, reaction, soft chemical, spray pyrolysis, electrospinning) with adequate figures of merit suitable for industrial applications (specially actuators, multi-layer ceramic capacitors, energy storage/harvesting, and dye degradation).

1 Introduction

1.1 INTRODUCTION TO GREEN ENERGY HARVESTING

Energy harvesting technologies have been the crux of modern-day miniaturized electronic devices for the last two decades. In a broader sense, energy harvesting necessitates capturing ambient forms of energy that are already present in the environment or device and converting them into electrical energy. Ambient energy including mechanical vibrations, heat, fluid flow, solar radiations, wind, waves and, most importantly, in vivo energies are capable of powering mobile electronics, wireless sensor networks, and wearable and implantable biomedical tools. Such devices play a very significant role in present as well as future advancements in industrial automation, public safety, energy management, and health care. To date, electrochemical batteries have been conventionally used as power sources to operate these devices because of their ease of installation. Compared to the life of electronic devices, batteries have a limited lifetime and hence require periodic replacement. Also, this incurs an extra recharging/replacement cost. Nevertheless, frequent replacement of batteries may lead to serious consequences in the case of biomedical implants by increasing the risk of infections and morbidity in patients. Yet another disadvantage of the batteries is their bulky nature, which dominates the size and weight of the electronic devices restricting their miniaturization. To overcome the above-discussed drawbacks, notable efforts have been taken up in research and development by not only the scientific community but also by industries in the direction of ambient energy harvesting technologies for self-powering a wide range of wireless electronic devices. Different energy harvesting technologies include electrostatic, electromagnetic, thermoelectric, triboelectric, piezoelectric, and pyroelectric transduction mechanisms at the micro, meso, and nanoscale. These technologies pose a positive socio-economic impact on current civilization. It is obvious from the exponential growth of the global energy harvester market, which is anticipated to increase from USD 484 million in 2020 to USD 800 million by 2025 with a CAGR of 10.6 percent (www.researchandmark ets.com/reports/5312292/global-energy-harvesting-system-market-2020), and also generate ample job opportunities. As per a research report, the business innovation observatory of the European Commission (https://ec.europa.eu), success stories of

DOI: 10.1201/9781003317289-1

1

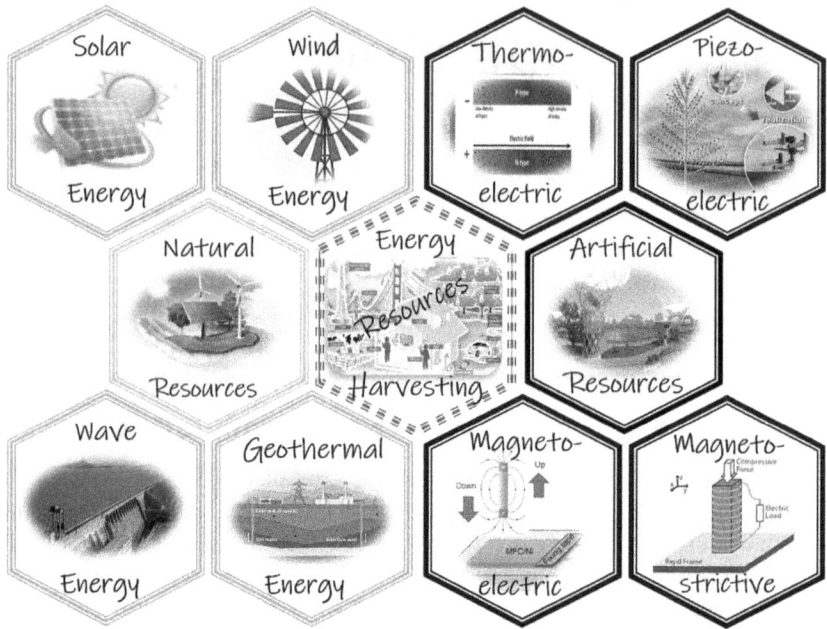

FIGURE 1.1 Schematic diagram showing the different natural and artificial energy resources.

a few companies patenting such technologies are listed in Table 1.1. Hence in this chapter, we discuss the renewable natural resources that are available in abundance and the ways to harness energy from them by employing different approaches, such as vibration/piezoelectric, thermoelectric, magnetoelectric, and magnetostrictive techniques.

1.2 ENERGY RESOURCES

Energy harvesting resources can be natural and artificial. Natural sources of energy include solar, wind, waves, and geothermal, while the artificial sources may be vibration/piezoelectric, magnetoelectric/magnetostrictive, thermoelectric, and so forth (Figure 1.1). Though the resources existing in nature are plentiful, energy can only be harnessed from them with artificial aids/means. In the current chapter, we will discuss energy harvesting from different natural and artificial resources in brief.

1.3 NATURAL SOURCES

1.3.1 SOLAR ENERGY

Solar energy is one of the best green energy resources available to mankind and is most viable and reasonably low-cost. It is simply the energy produced directly from the sun through thermonuclear reactions. The dosage of the sun's energy received by the earth in one and half hour is massive enough to handle the global energy needs for 24 hours. Apart from that, the sun's energy challenges the long-term threat of climate change posed by CO_2 emissions, one of the largest contributors to the greenhouse effect currently. Despite that, it accounts for only 1 percent of the electricity generation all over the world. This is mainly due to the intermittent nature of sunlight, which unpredictably fluctuates over time. Researchers around the globe are giving their sincere efforts to extract maximum solar power by employing different technologies. As per a report from Massachusetts Institute of Technology (MIT), photovoltaics (PV) is the most dominant technology in harnessing solar energy in the current scenario. A PV/ solar cell mostly consists of a PN junction. On the incidence of sunlight, charges start flowing, which constitutes an electric current when connected to load [1]. The first modern photovoltaic cell was fabricated in 1954 and stationed on a US satellite. After that, several initiatives have been carried out by the research community to implement large-scale use of solar cells to cater to energy demands but the inadequate efficiencies and high cost of the cells constitute the main hindrance. The best possible efficiency achieved with commercial-scale PV cells made out of Si single crystal is approximately 18 percent [2]. These so-called first-generation devices were overpriced in terms of fabrication and installation. Solar cells made out of $CuInGaSe_2$ polycrystalline semiconductor films (second generation) could reduce the cost but failed to improve the efficiency. The third generation emerged with the evolution of solar cells from thin films to dye-sensitized solar cells (DSSCs), organic cells, and bulk heterojunctions [3–5], which are much better in cost-effectiveness and large-scale energy conversion, but their efficiency remains less than 10 percent. Employing nanostructured semiconductors, and molecular and organic-inorganic hybrid assemblies to tune these third-generation solar cells to develop efficient photoanodes might be one of the solutions for increasing their efficiency. Brown et al. [6] incorporated titanium dioxide nanoparticles on highly conducting single-walled nanotubes (SWCNTs), which were later sensitized by $Ru(II)(bpy)^2(dcbpy)$. Though the SWCNT/TiO_2 film does not influence the charge injection but improves the charge separation. Incident photon to charge carrier efficiency was increased by a factor of 1.4 as a consequence of the SWCNT scaffold. This dye-sensitized SWCNT/TiO_2 cell exhibited an efficiency of 13 percent, open-circuited voltage (V_{oc}) of 0.26 V, and short circuit current density (J_{sc}) of 1.8 mA/cm². Besides, harvesting the near-infrared (NIR) light was a challenging trait in most of the DSSCs. In this regard, a strategy proposed by Hao et al. [7] for upconversion of non-responsive NIR light over a broader spectral range (190 nm, 670–860 nm) to a narrower solar cell responsive visible emission through dye-sensitized upconversion nanoparticle (DSUCNPs). They have used IR783-sensitized $NaYF_4$:10%Yb,2%Er@NaYF4:30%Nd core-shell $DSUCNP_s$ for enabling broadband NIR light-harvesting along with a spectral conversion into a narrow range of visible light. The sunlight-absorbing dyes (N719) in normal dye-sensitized solar cells can utilize only the ultraviolet-visible range. However, the introduction of DSUCNPs allows absorption in the

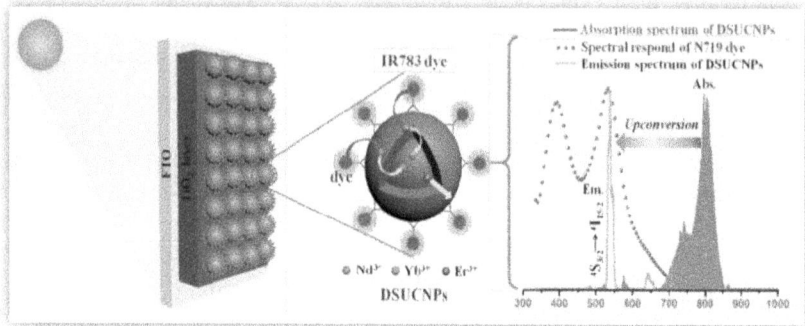

FIGURE 1.2   Working principle of broadband NIR solar light harvesting: (Left) Structure of DSSC with DSUCNPs placed on top of mesoporous TiO_2 layer; (Right) Activation of N719 by IR783 dye to enable spectral upconversion into the visible range.

Source: [7].

range of 670–860 nm. This is the NIR light, which constitutes almost 49 percent of solar radiations and Figure 1.2 shows the working principle of the device. There has been a remarkable development in photoelectric technology over the last five years. Very recently, a tandem organic solar cell comprising of electron beam evaporated $TiO_{1.76}$ and PEDOT: PSS interlayers across a unified direction demonstrates a power efficiency over 20 percent [8]. The future may foresee much greater advancements in this direction.

1.3.2   WIND ENERGY

Wind energy though an indirect form of solar energy is established as an essential member of the renewable energy club and can play a very major and important role in reducing dependence on fossil fuels. The wind is produced due to the differential heating of the earth's surface from the sun's energy. It is roughly estimated that 10 million megawatts of power can be available from wind, which may support the long-term sustainability of the global economy. The utilization of wind energy can be dated back to centuries ago, when the Chinese people used it for sailing boats, flying kites, and also constructed windmills for grinding grain, pumping water, and so forth. This form of energy is ubiquitous and prevalent all around in our surroundings, including natural wind in exterior space, flow in indoor heating, and ventilation air conditioning (AC) systems. Hence, apart from large-scale utilization of wind energy, it can also be employed to drive small self-powered wireless sensor networks. Generally, in wind energy harvesting systems, when a particular structure is subjected to the flow of air, limited cycle oscillations are set up in the system owing to structure-fluid interactions. This oscillation strain energy can be extracted into electricity through electromagnetic, electrostatic, and piezoelectric conversions [9]. Energy conversion on a small scale is the need of the hour for a smart lifestyle in present times. Research in this direction is experiencing substantial growth and has reported studies to enlist wind energy harvesting from miniaturized windmills, aero-elastic flutters, vortex, and turbulence-induced vibration, galloping, and so forth. Rancourt et al. [10] and eventually, Bansal et al. [11] tested miniature windmills on a centimeter scale. Kishore et al. [12] fabricated a mini-scale portable wind turbine of diameter 39.4 cm, which could produce power of 830 mW even at 5 m/s wind speed. Vortex-induced vibration (VIV) is another extensively studied methodology for fluid-based energy harvesting. In this technique, periodic amplitudes are created in the structure owing to repeated crosswind forces produced by the shedding of vortices in the wake of the given body. Pobering and Schwesinger [13] applied a prototype of an eel-like VIV harvester to extract energy from airflow and could roughly obtain energy density around 256 W/m^2 at an airflow speed of 10 m/s. Subsequently, harvesters based on flutter type design also became prevalent. Schmidt et al. [14] patented a two-oscillating blade-type flutter harvester with piezoelectric transduction. In a later study, it was theoretically estimated that 100 W/cm^3 of piezo-material could be derived from this design [15]. Another technique, galloping is usually used in scattered wind energy harvesting for application over all regions and seasons for its fantastic environmental adaptability. In this regard, Tan et al. [16] explored the blended influence of ambient temperature and wind speed on the performance of a galloping wind energy harvester with piezoelectric coupling. Outstanding adaptability to the environment was obtained in the range 40 °C–50 °C and a breeze to gale wind speed of 1–24 m/s. Apart from that, scientists are still trying to find out how to frame new techniques and methods to achieve maximum efficiency with better environmental adaptability. Recently, a unique wind-driven triboelectric nanogenerator (WTENG) based on PVDF and biodegradable leaf powder modified by poly-l-lysine (PLL) was designed for harvesting the mechanical energy of wind. The WTENG exhibits a marked increase in the I_{sc} values with the increase in wind speed and could go up to 150 μA at 10 m/s winds under test conditions. The power generated was large enough to drive the 'EXIT' led light for the exit passageway in windy weather [17]. The detailed design of the device is displayed in Figure 1.3.

1.3.3 WAVE ENERGY

Water bodies cover almost 70 percent of the earth's surface, ascertaining the enormous potential of wave energy to be channelized. Waves are usually generated when the wind blows over the surface of oceans or seas. Rigorous vertical movement of the ocean waves possess immense kinetic energy that can be captured by viable technologies to perform many useful tasks. It can cater to the high energy demands of marine facilities, including desalination devices, marine platforms, radar and other communication devices, weather forecasting, maritime transportation, exploration of deep-sea resources, monitoring of offshore structures, and electric charging stations located in the offshore area [18–19]. Oceans and seas also

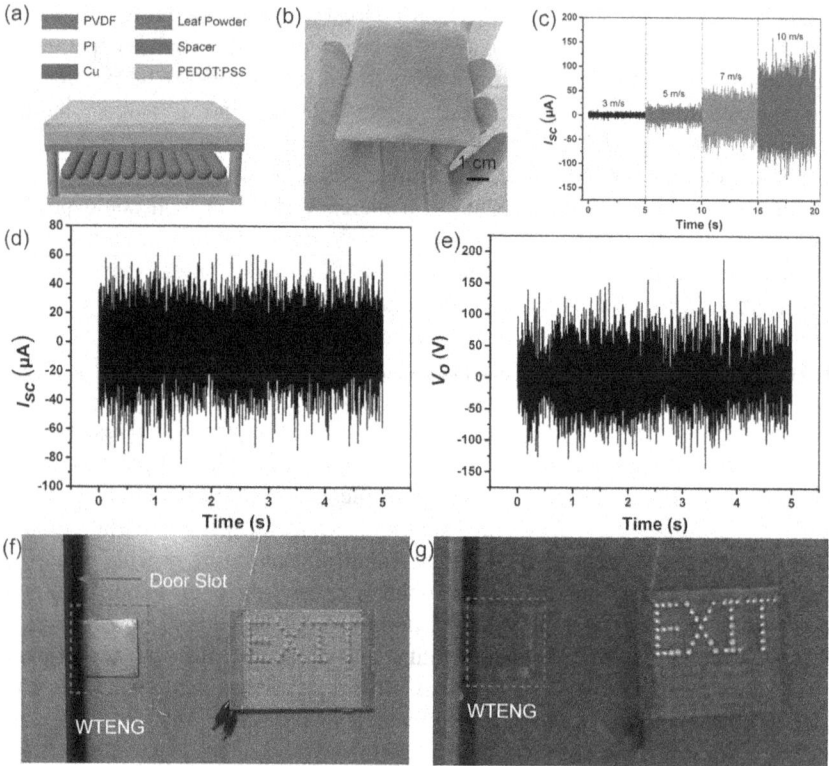

FIGURE 1.3 (a) Schematic diagram of WTENG based on leaf powder; (b) Actual picture of the WTENG; (c) I_{sc} values under different wind speeds; (d) and (e) Isc and Vo values at 7 m/s wind speed respectively; (f) and (g) Actual picture of LED scroller powered by WTENG triggered via wind through door slot.

Source: [17].

embrace thousands of moored and drifting buoys, floaters, and other information-collecting devices that integrate several monitoring sensors, telemetry equipment, a control system for ballast compensation, fishing feed, and so forth. All these devices require self-sustaining power sources that are difficult to procure in remote locations. However, with the surge of more and more ocean/sea-related activities, the energy demand is expected to show a manifold increase. Presently, almost all the offshore platforms maintain their battery packs powered by solar panels or small-scale wind turbines. But these well-established alternatives have a few drawbacks that may not be helpful in the long run. The use of solar power has issues related to its intermittent nature and low power yield (annual average ~170 W/m^2). Besides, the availability of wind energy is also sporadic though denser than solar power (~500 W/m^2 annually). Most importantly, exposure of the turbine blades to harsh maritime environmental conditions may adversely influence the dynamics of the floating platform or buoy. Therefore, cheaper, scalable, non-invasive, more reliable, and truly self-powered energy-harvesting technologies are essential to survive for long periods in massive water bodies or remote aquatic environments. Under such circumstances, harnessing wave energy in vivo is very essential. Aquatic energy sources are quite abundant, exhibiting an annual theoretical potential of ~151,300 TWh, among which ocean wave energy alone has a worldwide potential of about 18,500 TWh. Various technologies have been employed in the recent past to explore this energy, but there are still some relevant challenges to the practical deployment of wave energy harvesting techniques. Moreover, waves are irregular in terms of direction and frequency; and seasonal variations also affect the extracted energy. The marine environment in itself is the greatest hazard because of severe physical loads and biochemical degradation. In this regard, piezoelectric and triboelectric generators act as efficient and promising technologies due to their low-cost, high-power densities, and efficiencies. In 2001, Taylor et al. [20] employed eel-like equipment made up of flexible PVDF (polyvinyl fluoride) for converting the mechanical energy of ocean water to electrical energy. The generator used a regular trail of traveling vortices behind the bluff body to produce strain in the piezoelectric elements leading to undulating motions similar to that of an electric eel. With a novel idea, Murray et al. [21] constructed two-stage electric generators embedded in buoyant architectures that could interact with the ocean waves. Marine energy is harnessed via piezoelectric elements from a resonant secondary system. In another study, three rolling sphere-based encapsulated triboelectric nanogenerators (TENG) were placed inside the hull of a commercial model of a navigational buoy. The encapsulation helped to isolate the TENG from the harsh aquatic environment. In addition, this device demonstrated huge potential for harvesting low-frequency, small-amplitude waves [22]. Figure 1.4 depicts the electrical performance of different orientations of the TENG at peak-to-peak amplitude of 26° and pitch period of 1.486 s with the external load resistor varying from 100 Ω to 470 MΩ.

1.3.4 GEOTHERMAL ENERGY

Geothermal energy in its crude form arises from the temperature gradient existing between the tectonic plates in the earth's interior. This drives a continuous flow of thermal energy towards the earth's surface, which is relatively cooler. This source of energy is not only renewable and sustainable but also demonstrates the least dependence on good weather conditions. The common approaches to exploiting geothermal energy directly are a generation of electricity, satisfying energy requirements in buildings, agricultural and industries, melting of snow, and cultivation of plants. Since 1948, hot water/steam produced from geothermal energy has been utilized for snow melting in the United States. The earliest was structured in Klamath Falls, Oregon, and consists of a system of iron piping and a 50 percent ethylene glycol water solution circulating inside it [23]. In 2009, Ziegler et al. [24] configured airport aprons with a geothermal heating system to

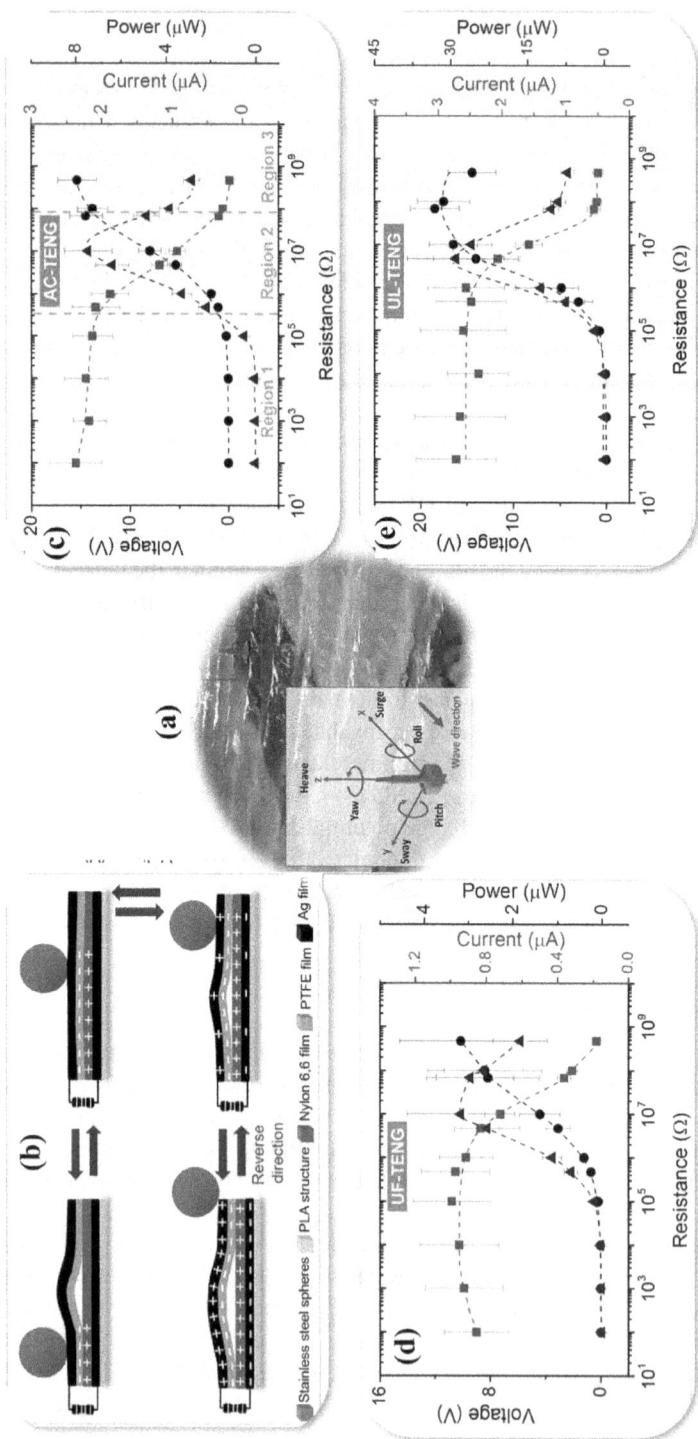

FIGURE 1.4   (a) Navigational buoy in wave basin; (b) Working technique of AC-TENG for harvesting wave energy; (c), (d) and (e) output voltage, current and power of AC-TENG, UF-TENG and UL-TENG respectively at a pitch amplitude of 26° and fixed period of 1.486 sec (AC-TENG, UF-TENG and UL-TENG are anisotropic circular based TENG, unidirectional flat-based TENG and unidirectional lateral based TENG respectively).

Source: [22].

remove snow at Greater Binghamton Airport, Johnson City, New York. Though the installation and maintenance costs were slightly more compared to traditional snow-removal methods, it was relatively safe and environment-friendly. Apart from that, geothermal power plants employ steam turbines to extract electricity from thermal energy. Recently, near-surface geothermal energy harvesting is gaining importance since it is capable of partially catering to the energy demands of residential as well as commercial buildings. A series of ground heat exchangers (GHE) is perpetually used to meet the heating and cooling exigencies of multi-stored large building complexes in various parts of the globe. Various forms of GHEs such as open and closed-loop systems with vertical and horizontal loops are used for this purpose, depending on their suitability of climate, local geology, heat exchange demand, and applications. Geothermal boreholes and geothermal piles are two familiar forms of heat exchangers. Piles with in-built GHE loops were efficiently installed in building foundations in countries like Germany, Austria, Switzerland, Australia, China, Japan, and so forth [25-27]. Tiwary et al. [28] modeled a pile-soil heat exchange to quantify the thermal interaction among the tiles and its role in minimizing the power output predicted from a series of geothermal piles. Finite elemental analysis of a pair of geothermal piles demonstrates the effects of spacing, diameter, thermal operation time, and orientation of embedded fluid circulation tubes (Figure 1.5). They could successfully predict the energy-harvesting efficiency of these geothermal piles that are thermally interacting. In another work, O. Ghasemi-Fare and P. Basu [29] made a comparative study between theoretical and experimental results on the heat transfer mechanism of piles in dry and saturated soil conditions. Recorded soil temperature data during the test cycle were compared to that of the predicted thermal response of the numerical model on pile-soil heat exchange. Laboratory tests of pore pressure transducers during thermal loading reveal the feasibility of temperature-induced pore fluid flow and convective heat transfer in saturated soil surrounding a ground heat exchanger. Hence, geothermal energy has a great potential that needs to be explored using more efficient techniques on a large scale.

1.4   ARTIFICIAL RESOURCES

From the above discussion, it is quite clear that ambient energy from the sun, wind, geothermal, waves, mechanical, and so forth be converted conveniently into electrical energy using exclusive techniques such as solar panels, piezoelectric, thermoelectric, electrostatic, magnetostrictive, magnetoelectric, electromagnetic, triboelectric generators, and so forth. We will discuss a few of the essential technologies here.

1.4.1   PIEZOELECTRIC

Piezoelectric energy harvesting technology deals with the ability of a class of materials to convert mechanical strain into electrical voltage. Strain may be

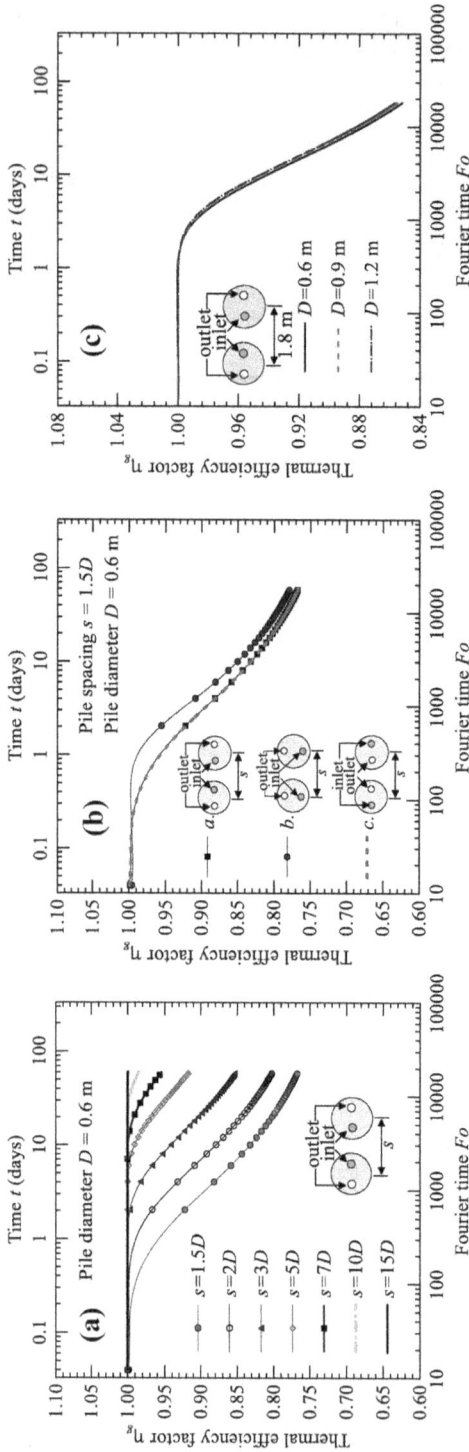

FIGURE 1.5 Effect of (a) pile spacing 'S' and duration of thermal operation 't' on the thermal efficiency of geothermal piles with diameter 'D' = 0.6 m in case of a two-pile group; (b) Orientations of fluid circulation tubes with pile spacing and diameter as 1.5D and diameter as 1.5D and 0.6 m respectively; (c) Pile diameter with spacing between the piles as 1.8 m on the thermal efficiency of piles.

Source: [28].

obtained from different sources, including human motion, acoustic noise, fluid flow, low-frequency seismic vibrations, and so forth when a piezoelectric material is strained; the dipoles present in material align themselves in such a way so as to generate a net polarization. This results in the development of electric potential across the crystal and is responsible for providing a convenient transducer effect. The extracted strain energy from these sources can be used to power many small-sized, portable, remote devices with low energy requirements, such as wearable electronics, implantable biomedical devices, sensors, MEMs (micromechanical systems), and global positioning system (GPS) receivers. One of the examples is the development of four proof-of-concept Heel strike generators by C.A. Howells [30] to transform the mechanical energy while walking into electrical energy. As the user walks, each heel strike of the boot generates power of almost 0.5 W at a step rate of 1 Hz which may be the efficient power source for many miniaturized devices like GPS receivers and communicators. As such, piezoelectric energy harvesting is quite advantageous for capturing energy from the surroundings, since it is entirely dependent on the intrinsic polarization of the piezoelectric crystal and there is no requirement for any external voltage source, magnetic field, or contact surfaces as seen in electromagnetic, electrostatic, and triboelectric energy harvesting systems. These are reliable, tough, highly sensitive, and yield approximately 3–5 fold higher power density and output voltage as compared to the other alternative harvesting methods. Hence not only small devices, but PEHs (piezoelectric energy harvesters) also find a place in structures, transportation, and IoT (Internet of things). Snyder [31] patented his idea of installing piezoelectric generators in car wheels to power tire pressure sensors. The harvesters gather strain energy from the wheel vibration during driving and can report abnormal tyre pressure to the driver via a radio link. Piezo harvesters can also derive in vivo energies from heartbeat, the motion of lungs, muscular bending and stretching, and the flow of blood to power biomedical instruments such as pacemakers, brain simulators, hearing aids, and sensors for measuring blood pressure, diagnosing heart rate and many more. Apart from that, day-to-day physical activities – for example footfalls, finger tapping, and hand swings – may be employed to power LEDs, cell phones, wristwatches, and so forth. In this regard, Alam and Mandal (2016) **[32]** fabricated a hybrid flexible piezoelectric generator (HPG) using native cellulose microfiber and polydimethylsiloxane along with multi-walled carbon nanotubes as conducting filler. It could deliver a power density of $9.0\ \mu W/cm^3$ with an open-circuited voltage of 30 V under repeated hand punching. This electrical throughput was large enough to power several LEDs, portable electronic units like commercial LCD screens, wristwatches, and calculators. Again, due to the biocompatibility of the HPG, this device may be suggested for biomedical applications as an implantable power source. Similarly, based on the piezoelectric principle, a self-powered implantable blood pressure monitoring device was developed by Cheng et al. [33] employing a polarized polyvinylidene fluoride (PVDF) to take a note of the hypertension status. This device delivered a maximum in vitro power of 2.3 μW and in vivo output power of 40 nW. Apart from that, suitable linearity in peak output voltage and pressure of the flow of blood, high sensitivity, and stability under iterating operating cycles revealed the potential of the device in clinical applications. Figure 1.6 exhibits the in vivo demonstration of the device wound on the ascending aorta of porcine. The response of the device to increasing blood pressure above 140 mm Hg is also displayed in the figure. The different aspects of piezoelectric energy harvesting will be covered in the following chapters.

1.4.2 THERMOELECTRIC

The thermoelectric (TE) effect is a physical phenomenon in which a temperature gradient can be converted into electrical voltage using a thermocouple. Generally,

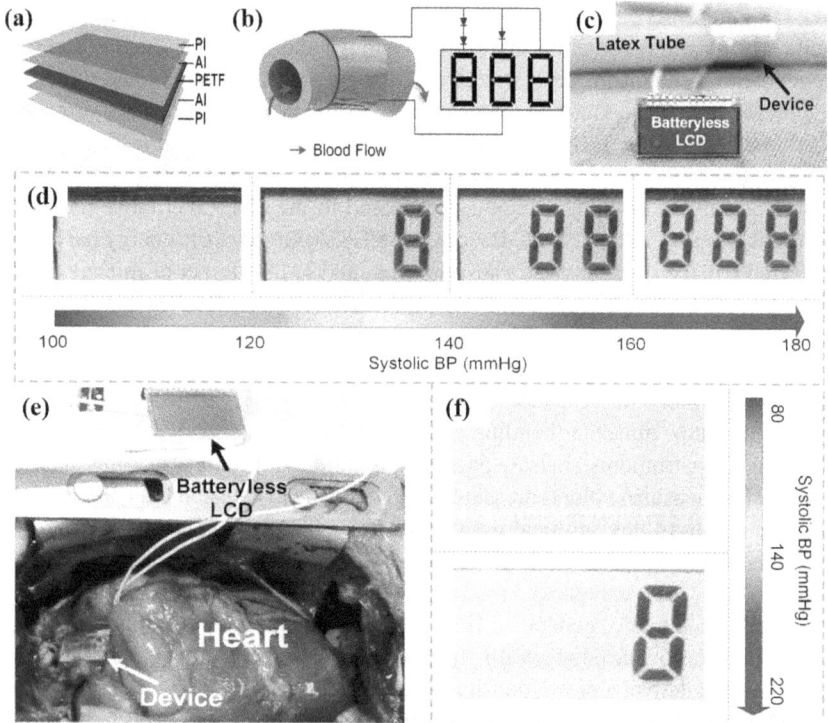

FIGURE 1.6 (a) Components of the implantable, self-powered blood pressure monitoring system; (b) Real picture of the system in vitro with the device wound around a latex tube and connected to the battery-less LCD; (c) Change in the LCD reading with an increase in flow pressure (FP); (d) Real-time photograph of the device is wrapped on the aorta and the battery-less LCD fixed on the thoracic wall of porcine; (e) LCD turned on with a systolic BP more than 140 mm Hg.

Source: [33].

TE technology is mainly based on the Seebeck effect apart from Peltier and Thompson effects. In the Seebeck effect, the drift of the charge carriers is driven by temperature differences across ends of thermoelectric materials (in conductors and semiconductors). Due to an existing temperature gradient at the junction of electron-deficient material (p-type) and electron-rich material (n-type), holes on the p-side transit to the n-side while electrons flow from n-side to p-side creating an electromotive force (emf). This emf is proportional to the temperature gradient between the hot and the cold ends. Seebeck effect is characterized by the Seebeck coefficient $\alpha \equiv \dfrac{V}{\Delta T}$; where V is the potential difference (or emf) and ΔT is the temperature difference between the ends of two dissimilar materials (p-type and n-type).

Thermoelectric generators (TEGs) that are based on the Seebeck effect are usually exploited to convert thermal energy into electrical energy. The earliest use of TEG can be dated back to 1961, when it was used by NASA to power the Transit navigation satellite. The chief advantages of this technology are its simplicity in principle (does not require complicated arrangement), longer lifespan, workability in extreme environments such as high temperature, high voltage output, biocompatibility for powering medical implants, and it creates less noise. Space missions require wireless sensors for orbiting satellites and land vehicles that have to face drastic thermal environments. TEGs could be thought of to be an alternative power source under such circumstances. Various types of thermoelectric generators have been proposed in the past. However, Tu et al. [34] suggested a thermoelectric energy harvester using a phase change material (paraffin-based composite + expanded graphite) under extreme temperature variations of +100 to–50 °C for application in space. The performance of the device was evaluated both theoretically and experimentally. Experimental evidence reveals that 5 wt percent expanded graphite/paraffin composite-based TEG demonstrates the highest total energy output. Body energy harvesting is another potential alternative to batteries to improve the functionality of wearable devices. A substantial amount of energy in the human body is released in the form of motion and heat, out of which mechanical efficiency is only about 15–30 percent [35]. The remaining part of the energy supplied by food is released as body heat, which can be utilized as a continuous source of energy (approx. 60–180 W depending on activity). There is a possibility that if the thermoelectric devices capture energy with a conversion efficiency of ~1 percent, the power generated would amount to ~0.6–1.8 W, large enough to charge wearable electronics.

The Seiko Thermic watch, assembled in 1998, was the first-ever commercial body heat power wrist watch [36]. However, the thermal conversion efficiency was only 0.1 percent with its open-circuited voltage (V_{oc}) of 300 mV and output power of 25 µW for a 1.5 °C difference between the thermoelectric modules. Sometime later, another TEG-powered wristwatch was designed by Citizen Watch Co. in 1999 [37] with $V_{oc} \sim$ 640 mV and power ~ 13.8 µW. In the recent past, PowerWatch, a smartwatch that draws all its power from body heat, was successfully developed by Matrix Industries [38]. The thermoelectric efficiency of a newly modeled wearable and flexible TEG was tested for its functionality using the finite volume method and also experimentally (Figure 1.7) by Wang et al. in 2017 [39]. The model uses a thermal interface layer to extract heat from human skin. The results obtained in this work could be employed for optimal structural designs for wearable TEGs and material selection for the interface layer to improve power generation for wearable electronics. In one of the newest studies, successful utilization of vertical temperature gradient was demonstrated by single-walled carbon nanotube films by a three-dimensional spring-shaped novel design structure. The device was compressible and flexible enough to underpin interfacial contact for efficient heat transfer from skin to the device's hot end while incorporated air gaps (thermally insulating) limited the vertical heat transfer internally within the device for maintenance of a much better temperature gradient. Consequently, 3 pairs of p-n couples with air gaps generated reasonable output power of 749.19 nW at a temperature difference of 30 K [40]. Similarly, thermoelectric energy harvesting can be availed from asphalt pavements to not only reduce the temperature of the roads to lessen the damage caused by high temperatures, but also employed to light LED lamps, in-suit monitoring sensors, and much more [41].

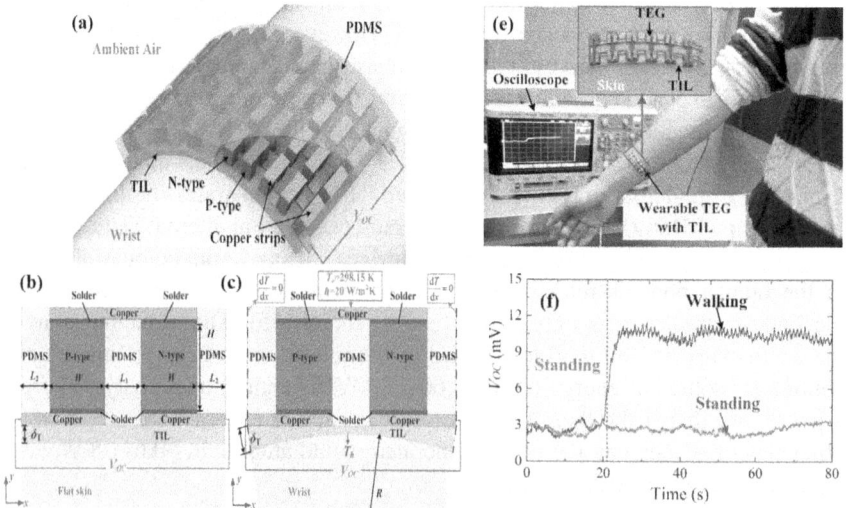

FIGURE 1.7 (a) Schematic diagram showing the constructional design of a thermoelectric generator placed on the wrist; Cross-sectional view of the generator when (b) the skin is acting like a flat surface and (c) the skin is acting like a curved surface; (e) Experimental demonstration of the wearable device; (f) Measured Voc values with time while standing and walking with arm swing.

Source: [39].

1.4.3 MAGNETOSTRICTIVE

The magnetostrictive effect is another green alternative source for harvesting vibration energy. This effect was first observed by James Prescott Joule in 1842 when he found a change in length with the application of a magnetic field. Later, Villari suggested a change in the magnetization of a special class of materials (magnetostrictive) for applying stress. This category of magneto–mechanical coupling could be discussed in terms of the Stoner-Wohlfarth approximation, which considers a magnetostrictive material as an assemblage of non-interacting magnetic domains exhibiting uniform local magnetization M_s. Total magnetization of the material can be expressed as the weighted sum of local response of the magnetic domains towards stress and a magnetic field. Under the dominance of mechanical compression, magnetic domains are pinned in a direction perpendicular to the direction of stress, giving rise to zero net magnetization. However, under the influence of the magnetic field, the domains become oriented parallel to the direction of the field, resulting in maximum net magnetization. Maximum magneto-mechanical coupling corresponds to the perpendicular alignment of the magnetic domains. Under this condition, magnetic energy is perfectly balanced with mechanical energy. Mechanical stress in magnetostrictive materials is applied directly by vibration sources or derived indirectly from vibration-induced inertial force. Besides, the bias magnetic field is provided mostly by permanent magnets to balance the stress. The constitutive relations for magneto-mechanical coupling, small coaxial stress and magnetic field perturbations are given by: $\Delta B = d\Delta T + \mu^H \Delta H$ and $\Delta S = s^H \Delta T + d\Delta H$; where s^H and μ^H are the elastic compliance and magnetic permeability respectively while ΔB, ΔS, ΔT and ΔH are the change in magnetic flux density, strain along the direction of input, stress and magnetic field intensity respectively. magnetostrictive energy harvesters are inductive due to which they can provide low impedance at fundamental frequencies of common structural vibration sources.

Terfenol-D, a rare earth alloy of Tb, Dy, and Fe ($Tb_{0.3}Dy_{0.7}Fe_2$), exhibits the highest magnetostriction at room temperature and is presently being used for many applications. Other compounds exhibiting sizable magnetostriction are Fe-Ga alloy (Galfenol) and Metglas [42]. Recently, Yan et al. [43] proposed the design and fabrication of a magnetostrictive generator based on the rotary impact of a human knee joint. It is constructed out of 12 movable Terfenol-D rods, with each one circumscribed by picked-up coils and alternate permanent magnet arrays squeezed into each part of the shell. A prototype of the harvester could generate a huge induced voltage of 60–80 volts at a very low-frequency rotation of 0.91 Hz. The device also demonstrated good performance levels at low frequencies of human walking and periodic swing crus situation suggesting its application in wearable knee joint applications. Similarly, a novel magnetostrictive power generator employing a Fe-Ga plate was proposed for battery-free IoT application by T. Ueno [44]. The prototype device (4 gm in weight) with dimensions 4 x 0.5 × 16 mm of Fe-Ga plate yield a $V_{oc} \sim 4$ V and effective power of 2.0 mW at 88.7 Hz and 6.0 m/

s^2. The harvester performance was good enough to substitute for the button cells in wireless modules.

1.4.4 Magnetoelectric

The Magnetoelectric (ME) effect usually deals with coupling between magnetic energy and electrical energy. On application of a magnetic field, the ME transducer produces a change in electric polarization (direct ME effect) while magnetization variations are observed as a response to an applied electric field (indirect ME effect). The two-way functionality of the magnetoelectric materials makes them suitable for multifarious applications such as sensor, actuator, and energy harvester applications. Though the ME effect is observed in single-phase materials such as $BiFeO_3$ and Cr_2O_3, composites fabricated out of magnetostrictive and piezoelectric layers are usually preferred for energy harvesting applications. Magnetoelectric generators can be employed for harnessing energy in mobile base stations, satellite communications, Wi-Fi routers, television and radio transmitters, power distribution lines, and so forth. In this regard, Silva et al. [45] constructed bilayered and trilayered flexible ME composites of varying Vitrovac and PVDF. They realized a more prominent magnetoelectric behavior (75 V/cmOe) with the trilayered structure (magnetostrictive/piezoelectric/magnetostrictive laminate) than with the bilayer configuration (66 V/cmOe). A flexible and low-cost energy harvesting device was designed by Lasheras et al. [46] in 2015, using amorphous magnetostrictive $Fe_{64}Co_{17}Si_7B_{12}$ ribbons and PVDF piezoelectric element. This 3 cm long sandwiched structure could deliver a power output of 6.4 µW and power density of 1.5 mW cm^{-3} at optimum load resistance measured at magnetomechanical resonance of the laminate. In a similar report, Dai et al. [47] proposed a broadband vibration energy harvester engaging a rotary pendulum and ME transducer. The harvester displayed a 3db bandwidth of 3.22 Hz at 0.5 g RMS acceleration. Besides, the device has a frequency doubling feature which religiously increased the output power at lower frequencies and 970.2 µW power output at 0.5g RMS acceleration was delivered by the device at 14.8 Hz (resonant frequency). Recently, an approach to design a trilayer magnetoelectric composite using Metglas/PVDF/Metglas was proposed that generated a power output and power density of 12 µW and 0.9 mW cm^{-3} respectively. Such a notable performance was particularly useful for microdevice applications in hard-to-reach locations and traditional devices, including door locking, electric windows, tire pressure monitoring, and so forth [48]. The above discussions mirror the depth of research in the direction of ME energy harvesting, though this field is still in its infancy and needs much attention.

1.5 SUMMARY

In the present scenario, stand-alone systems such as wireless sensor networks in remote locations, implantable biomedical devices, and wearable and mobile electronics are becoming increasingly pervasive. Under such circumstances, harvesting energy from the ambient environment is one the finest solutions to power these self-sustainable gadgets. Energy can be scavenged from both available natural and artificial resources. This chapter discusses the harvesting possibilities and challenges associated with different types of natural sources such as solar, wind, waves, and geothermal; and artificial sources such as piezoelectric, thermoelectric, magnetostrictive, and magnetoelectric. The probability of all-around dependence on ambient resources will strike out the hindrances associated with the limited reliability of standard batteries. Apart from that, it will also extend the lifespan of the sophisticated gadgets, support the conventional electronic systems, and also reduce the maintenance cost. Hence, the use of readily available energy in the surroundings will satisfy our goal of making this earth an eco-friendly and smarter place to live.

REFERENCES

1. Wang, H., Jasim, A. and Chen, X., 2018. Energy harvesting technologies in roadway and bridge for different applications–A comprehensive review. *Applied Energy*, *212*, pp. 1083–1094.
2. Crabtree, G.W. and Lewis, N.S., 2007. Solar energy conversion. *Physics Today*, *60*(3), pp. 37–42.
3. O'regan, B. and Grätzel, M., 1991. A low-cost, high-efficiency solar cell based on dye-sensitized colloidal TiO_2 films. *Nature*, *353*(6346), pp. 737–740.
4. Van De Lagemaat, J., Barnes, T.M., Rumbles, G., Shaheen, S.E., Coutts, T.J., Weeks, C., Levitsky, I., Peltola, J. and Glatkowski, P., 2006. Organic solar cells with carbon nanotubes replacing In_2O_3: Sn as the transparent electrode. *Applied Physics Letters*, *88*(23), p. 233503.
5. Yu, G., Gao, J., Hummelen, J.C., Wudl, F. and Heeger, A.J., 1995. Polymer photovoltaic cells: Enhanced efficiencies via a network of internal donor-acceptor heterojunctions. *Science*, *270*(5243), pp. 1789–1791.
6. Brown, P., Takechi, K. and Kamat, P.V., 2008. Single-walled carbon nanotube scaffolds for dye-sensitized solar cells. *The Journal of Physical Chemistry C*, *112*(12), pp. 4776–4782.
7. Hao, S., Shang, Y., Li, D., Ågren, H., Yang, C. and Chen, G., 2017. Enhancing dye-sensitized solar cell efficiency through broadband near-infrared upconverting nanoparticles. *Nanoscale*, *9*(20), pp. 6711–6715.
8. Zheng, Z., Wang, J., Bi, P., Ren, J., Wang, Y., Yang, Y., Liu, X., Zhang, S., and Hou, J., 2022. Tandem organic solar cell with 20.2% efficiency. *Joule*, *6*(1), pp. 171–184.
9. Zhao, L. and Yang, Y., 2017. Toward small-scale wind energy harvesting: Design, enhancement, performance comparison, and applicability. Shock and Vibration.
10. Rancourt, D., Tabesh, A. and Fréchette, L.G., 2007. Evaluation of centimeter-scale micro windmills: Aerodynamics and electromagnetic power generation. *Proc. Power MEMS*, *20079*, pp. 93–96.
11. Bansal, A., Howey, D.A., and Holmes, A.S., 2009, June. Cm-scale air turbine and generator for energy harvesting from low-speed flows. In *TRANSDUCERS*

2009-2009 International Solid-State Sensors, Actuators and Microsystems Conference (pp. 529–532). IEEE.

12. Kishore, R.A., Coudron, T. and Priya, S., 2013. Small-scale wind energy portable turbine (SWEPT). *Journal of Wind Engineering and Industrial Aerodynamics, 116,* pp. 21–31.

13. Pobering, S. and Schwesinger, N., 2008, October. Power supply for wireless sensor systems. In *SENSORS, 2008 IEEE* (pp. 685–688). IEEE.

14. Schmidt, V.H., 1985. *Piezoelectric Wind Generator.* U.S. Patent 4,536,674.

15. Schmidt, V.H., 1992, October. Piezoelectric energy conversion in windmills. In *IEEE 1992 Ultrasonics Symposium Proceedings*(pp. 897–904). IEEE.

16. Tan, T., Zuo, L. and Yan, Z., 2021. Environment coupled piezoelectric galloping wind energy harvesting. *Sensors and Actuators A: Physical, 323,* p. 112641.

17. Feng, Y., Zhang, L., Zheng, Y., Wang, D., Zhou, F., and Liu, W., 2019. Leaves-based triboelectric nanogenerator (TENG) and TENG tree for wind energy harvesting. *Nano Energy, 55,* pp. 260–268.

18. Viet, N.V., Xie, X.D., Liew, K.M., Banthia, N. and Wang, Q., 2016. Energy harvesting from ocean waves by a floating energy harvester. *Energy, 112,* pp. 1219–1226.

19. Rodrigues, C., Nunes, D., Clemente, D., Mathias, N., Correia, J.M., Rosa-Santos, P., Taveira-Pinto, F., Morais, T., Pereira, A. and Ventura, J., 2020. Emerging triboelectric nanogenerators for ocean wave energy harvesting: state of the art and future perspectives. *Energy & Environmental Science, 13*(9), pp. 2657–2683.

20. Taylor, G.W., Burns, J.R., Kammann, S.A., Powers, W.B. and Welsh, T.R., 2001. The energy harvesting eel: a small subsurface ocean/river power generator. *IEEE Journal of Oceanic Engineering, 26*(4), pp. 539–547.

21. Murray, R. and Rastegar, J., 2009, April. Novel two-stage piezoelectric-based ocean wave energy harvesters for moored or unmoored buoys. In *Active and Passive Smart Structures and Integrated Systems 2009* (Vol. 7288, pp. 184–195). SPIE.

22. Rodrigues, C., Ramos, M., Esteves, R., Correia, J., Clemente, D., Gonçalves, F., Mathias, N., Gomes, M., Silva, J., Duarte, C. and Morais, T., 2021. Integrated study of triboelectric nanogenerator for ocean wave energy harvesting: performance assessment in realistic sea conditions. *Nano Energy, 84,* p. 105890.

23. Lienau, P.J. and Culver, G., 1989. Klamath falls geothermal field, Oregon: a case history of assessment, development, and utilization. *P.J. Lienau, G. Culver, J.W. Lund. Sept. 1989, 123.*

24. Ziegler, W., 2009. Radiant heating of airport aprons. *Airport Operations and Maintenance Challenge. Binghamton University.*

25. Brandl, H., 2006. Energy foundations and other thermo-active ground structures. *Géotechnique, 56*(2), pp. 81–122.

26. Gao, J., Zhang, X., Liu, J., Li, K. and Yang, J., 2008. Numerical and experimental assessment of the thermal performance of vertical energy piles: an application. *Applied Energy, 85*(10), pp. 901–910.

27. McCartney, J.S., Sánchez, M. and Tomac, I., 2016. Energy geotechnics: Advances in subsurface energy recovery, storage, exchange, and waste management. *Computers and Geotechnics, 75,* pp. 244–256.

28. Tiwari, A.K., Kumar, A. and Basu, P., 2022. The influence of thermal interaction on energy harvesting efficiency of geothermal piles in a group. *Applied Thermal Engineering, 200,* p. 117673.

29. Ghasemi-Fare, O. and Basu, P., 2018. Influences of ground saturation and thermal boundary condition on energy harvesting using geothermal piles. *Energy and Buildings*, *165*, pp. 340–351.

30. Howells, C.A., 2009. Piezoelectric energy harvesting. *Energy Conversion and Management*, *50*(7), pp. 1847–1850.

31. Snyder, D.S., Imperial Clevite, 1983. *Vibrating transducer power supply for use in abnormal tire condition warning systems*. U.S. Patent 4,384,482.

32. Alam, M.M. and Mandal, D., 2016. Native cellulose microfiber-based hybrid piezoelectric generator for mechanical energy harvesting utility. *ACS Applied Materials & Interfaces*, *8*(3), pp. 1555–1558.

33. Cheng, X., Xue, X., Ma, Y., Han, M., Zhang, W., Xu, Z., Zhang, H. and Zhang, H., 2016. Implantable and self-powered blood pressure monitoring based on a piezoelectric thin film: Simulated, in vitro, and in vivo studies. *Nano Energy*, *22*, pp. 453–460.

34. Tu, Y., Zhu, W., Lu, T. and Deng, Y., 2017. A novel thermoelectric harvester based on high-performance phase change material for space application. *Applied Energy*, *206*, pp. 1194–1202.

35. Winter, D.A., 2009. *Biomechanics and Motor Control of Human Movement*. John Wiley & Sons.

36. Qi, Y. and McAlpine, M.C., 2010. Nanotechnology-enabled flexible and biocompatible energy harvesting. *Energy & Environmental Science*, *3*(9), pp. 1275–1285.

37. Shinohara, Y., 2017. Recent progress of thermoelectric devices or modules in Japan. *Materials Today: Proceedings*, *4*(12), pp. 12333–12342.

38. Matrix power watch. Powerwatch, Matrix Industries. www.powerwatch.com/ [accessed 27 Aug. 2018].

39. Wang, Y., Shi, Y., Mei, D. and Chen, Z., 2017. Wearable thermoelectric generator for harvesting heat on the curved human wrist. *Applied Energy*, *205*, pp. 710–719.

40. Lv, H., Liang, L., Zhang, Y., Deng, L., Chen, Z., Liu, Z., Wang, H. and Chen, G., 2021. Flexible spring-shaped architecture with optimized thermal design for wearable thermoelectric energy harvesting. *Nano Energy*, *88*, p.106260.

41. Zhu, X., Yu, Y. and Li, F., 2019. A review on thermoelectric energy harvesting from asphalt pavement: Configuration, performance, and future. *Construction and Building Materials*, *228*, p.116818.

42. Narita, F. and Fox, M., 2018. A review on piezoelectric, magnetostrictive, and magnetoelectric materials and device technologies for energy harvesting applications. *Advanced Engineering Materials*, *20*(5), p. 1700743.

43. Yan, B., Zhang, C. and Li, L., 2018. Magnetostrictive energy generator for harvesting the rotation of human knee joint. *AIP Advances*, *8*(5), p. 056730.

44. Ueno, T., 2019. Magnetostrictive vibrational power generator for battery-free IoT application. *AIP Advances*, *9*(3), p. 035018.

45. Silva, M.P., Martins, P., Lasheras, A., Gutiérrez, J., Barandiarán, J.M. and Lanceros-Mendez, S., 2015. Size effects on the magnetoelectric response on PVDF/Vitrovac 4040 laminate composites. *Journal of Magnetism and Magnetic Materials*, *377*, pp. 29–33.

46. Lasheras, A., Gutiérrez, J., Reis, S., Sousa, D., Silva, M., Martins, P., Lanceros-Mendez, S., Barandiarán, J.M., Shishkin, D.A. and Potapov, A.P., 2015. Energy

harvesting device based on a metallic glass/PVDF magnetoelectric laminated composite. *Smart Materials and Structures*, *24*(6), p. 065024.

47. Dai, X., 2016. A vibration energy harvester with broadband and frequency-doubling characteristics based on rotary pendulums. *Sensors and Actuators A: Physical*, *241*, pp. 161–168.

48. Reis, S., Silva, M.P., Castro, N., Correia, V., Rocha, J.G., Martins, P., Lasheras, A., Gutierrez, J. and Lanceros-Mendez, S., 2016. Electronic optimization for an energy harvesting system based on magnetoelectric Metglas/poly (vinylidene fluoride)/ Metglas composites. *Smart Materials and Structures*, *25*(8), p. 085028.

2 Piezoelectric Figure of Merits

2.1 FUNDAMENTALS OF PIEZOELECTRICITY

Piezoelectricity is a characteristic property exhibited by some materials in response to mechanical stimulation to generate electrical charge and vice versa. Piezoelectricity was first discovered in Rochelle salt in 1880 by French scientists Jacques and Pierre Curie and is exhibited by crystals without a center of symmetry (non-centrosymmetric structure). A detailed classification of the crystals according to their symmetry elements of translation and orientation about a point is given in Figure 2.1 [1]. The microscopic origin of the piezoelectric effect is the displacement of ionic charges within the crystal structure. Mechanical stress employed to such a specimen creates an overall polarization inducing a voltage change across it. On reversing the direction of stress, the direction of polarization reverses and, in turn, the voltage. Hence, the direct piezoelectric effect refers to the creation of electric polarization by mechanical stress (requisite for generator operation), whereas the converse of piezoelectric effect indicates the mechanical displacement triggered by an electric field (requisite for motor operation). Both direct and indirect piezoelectric effects are presented diagrammatically in Figure 2.2. In the first, the dipole moments (produced due to displacement of ionic charges) in the crystal just cancel each other under unstrained conditions based on the first mechanism. Also, in this case, electric polarization is linearly related to mechanical stress and is named as *linear piezoelectric effect*. However, based on the second mechanism, the dipole moments may remain and get added up to a resultant moment along the polar axis of the unit cell giving rise to piezoelectricity along with pyroelectricity. Figure 2.2 depicts some of the applications of the direct and indirect piezoelectric effect.

2.1.1 DOMAIN STRUCTURE AND DOMAIN WALL MOTION

Piezoelectric crystals are characterized by a well-defined domain structure. Domains are basically the regions within those crystals containing quite a substantial number of electric dipoles oriented in the same direction. These are regions of uniform polarization within the crystal that decrease the energy associated with

DOI: 10.1201/9781003317289-2

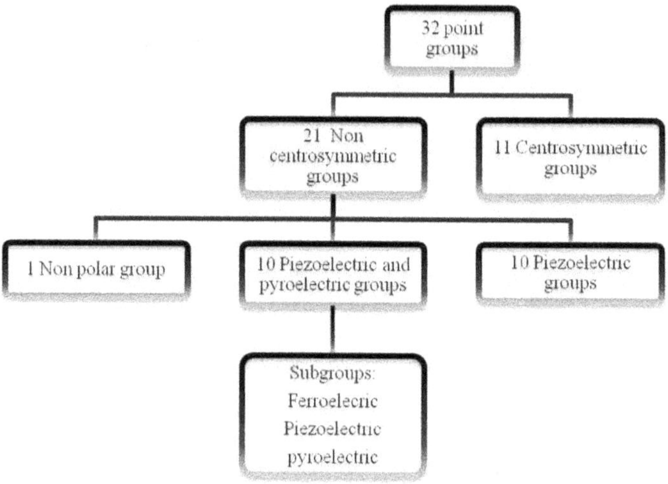

FIGURE 2.1 Flow chart showing the division of space groups.

Source: [1].

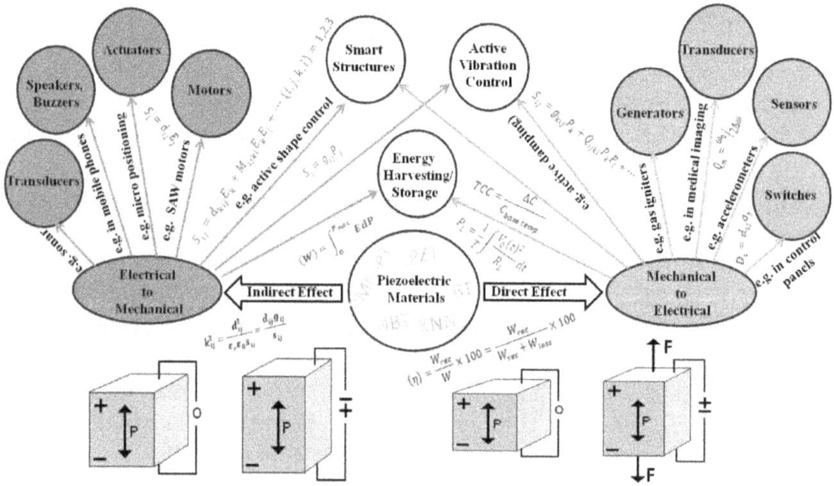

FIGURE 2.2 Schematic diagram showing the various applications of piezoelectric materials (the background shows different piezoceramics along with the material parameters essential for specific applications).

the depolarization field [2]. They are aligned in such a way that the polarization for each domain will compensate for the other and the net polarization is zero. The interfaces that separate the domains are called domain walls, which are the loci of dipole orientation from one direction of the domain to another of the neighboring ones. When a phase transition begins, the domains are nucleated at several regions within the polycrystalline material, and the nuclei of the domains preferably grow along the crystallographic axes until the transition to a new phase is completed throughout the volume. Ferroelectric domains closely resemble the ferromagnetic domains. In ferromagnets, the typical energy of a domain wall is primarily governed favoring parallel alignment of the spins and magnetic domain walls. Such walls usually represent a gradual rotation of spin over tens or hundreds of nanometers leading to the so-called Bloch or Neel type of domain walls (Figure 2.3(a)). On the other hand, the domain wall energy in ferroelectrics is dominated by a very strong coupling between polarization and strain and therefore limited to a shorter length, that is, of the order of a few unit cells. Such domain walls in ferroelectrics are of Ising type [3, 4]. Considering the case of uniaxial ferroelectrics such as TGS crystals, domains can have only two possible dipole orientations. Hence, the polarization will align in opposite directions in adjacent domains by twinning and are called 180° walls. However, in multiaxial ferroelectrics like $BaTiO_3$, the domains can possess more than two dipole orientations. For example, in the case of a tetragonal ferroelectric phase, the polarization can be along six easy directions deriving from the cubic structure, that is, the elongated c-axis ($\pm$ X, $\pm$ Y, and $\pm$ Z directions). This gives rise to different types of domain walls: those separating antiparallel dipoles are 180° domain walls while those separating dipoles at 90° (non-180°) walls. Similarly, for the ferroelectric phase with the rhombohedral unit cell, the polarization is along the body diagonals allowing 8 easy directions for spontaneous polarization with angles 70.5° and 110° walls. The polarization direction may differ by 180° or non-180° in different domains and one domain state can be switched into another by the application of the electric field [1].

2.1.2 P-E Hysteresis Loop

The effect of applied electric field on the domain structure leading to the polarization hysteresis loop is shown in Figure 2.3(b). The points A, B, C, D, and E are marked by dots in the graph. Application of an electric field E (or voltage V) to the crystal, results in the polarization of the domains parallel to the external field which increases at the cost of those domains having a different direction of polarization. Thus, the polarization P increases and its dependence on the electric field is presented by the curve B-C. The polarization reaches a saturation value P_s when all the domains align themselves parallel to the externally applied electric field. It may be said that the crystal has grown into a single domain. Extrapolation of the linear part at point C gives the value of spontaneous polarization and is equivalent to the polarization that existed in each domain in its inceptive state represented by point B in the figure. Hence, spontaneous polarization refers to the polarization of discrete domains and not the polarization of the crystal as a whole. Again, as the intensity of the electric field decreases, the polarization also reduces. However, even if the electric field intensity drops to zero, a remnant polarization P_r still remains within the crystal, which is depicted by the curve C-D. Further applying an electric field in the opposite direction, the dipoles tend to orient

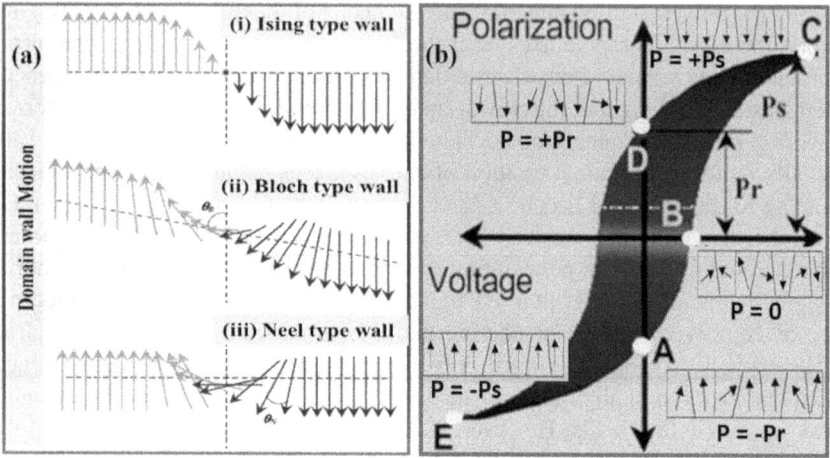

FIGURE 2.3 (a) Schematic representation of (i) Ising type (ii) Bloch type and (iii) Neel type of domain walls. The Bloch and Neel rotation are represented by θ_B and θ_N respectively; (b) orientation of domains along the hysteresis loop.

themselves in the reverse direction. As a result, a considerable portion of the crystal gets polarized in the opposite way, destroying the remnant polarization. This value of the electric field is known to be the coercive field. Consequently, the sample changes its sense of polarization in accordance with the alignment of the domains and corresponds to the new direction of the electrical field, eventually attaining its maximum value (-Ps), as represented in Figure 2.3(b). Reversal of polarization in the domains with the commutation in the direction of the applied electric field is said to be switching of domains. This hysteresis loop is generally observed within a certain temperature range demarcated by a transition temperature called the Curie temperature T_c. Usually, the crystal no longer behaves as a piezoelectric material for temperatures beyond T_c and exhibits normal paraelectric behavior (non-polar phase). The dielectric constant ε shows an abnormally high value in the neighborhood of this temperature. At $T > T_c$, this anomalous behavior follows the Curie–Weiss law:

$$\varepsilon = \frac{C}{T - T_c}$$

where C is called as the Curie constant. In fact, this kind of anomalous behavior can be seen whenever there is a phase transition even below T_c. Such anomalies include not only dielectric constant but also polarization, elastic and piezoelectric constants; and specific heat, since T_c corresponds to a change in the crystal structure.

2.2 CONSTITUTIVE RELATIONS

It is well known that the significant interest in piezoelectric materials is due to their inherent ability to convert mechanical vibrations into voltage and vice versa. Based on their electromechanical properties, a nomenclature of a set of figures of merit has been devised to assist in the selection of materials for particular applications. Hence, in this section, we have tried to organize the various key parameters affecting the performance of a device/material. A schematic diagram showing the piezoelectric figure of merit governing different applications is displayed in Figure 2.2. The piezoelectric properties of a material depend on the direction of orientation of the polar axis and are normally described in terms of tensors. Conventionally, subscripts i j k l are chosen to define the direction and orientation as illustrated in Figure 2.4. The subscript '3' refers to the direction of the polar axis (or poling axis); '1' and '2' refer to an arbitrary chosen orthogonal axis perpendicular to '3'. Similarly, the subscripts '4', '5', '6' refer to the shear planes of mechanical stress and strain perpendicular to axes '1', '2' and '3' respectively [1].

2.2.1 PIEZOELECTRIC CHARGE CONSTANT (D) AND PIEZOELECTRIC VOLTAGE CONSTANT (G)

In tensor notation, the electric field induced strain S_{ij} can be represented in power series either in terms of electric field E_k or polarization P_k as:

$$S_{ij} = d_{kij}E_k + M_{ijkl}E_kE_l + \ldots (i,j,k,l) = 1,2,3,4 \qquad (1)$$

$$S_{ij} = g_{kij}P_k + Q_{ijkl}P_kP_l + \ldots \qquad (2)$$

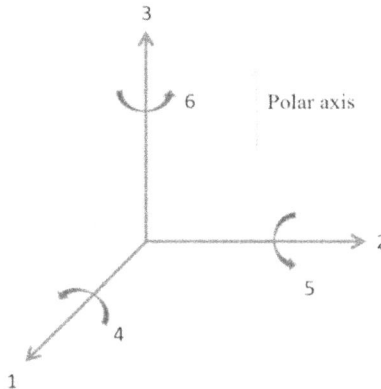

FIGURE 2.4 Subscript symbols for the notation of the directional properties for the poled piezoelectric ceramics.

Here d_{kij} and g_{kij} indicate the piezoelectric coefficients, whereas M_{ijkl} and Q_{ijkl} are the electrostrictive coefficients. In either of the equations, d and g have been isolated by naming them as piezoelectric charge/strain constant d and piezoelectric voltage constant g. The first term in both the equations stems from converse piezoelectric effect and the second term from electrostriction. It is often sufficient to consider the longitudinal and transverse displacements from the point of view of materials development. For small applied fields, the above relations (1) and (2) become simplified and maintain linearity as:

$$S_i = d_{ij}E_j \tag{3}$$

$$S_i = g_{ij}P_j \tag{4}$$

The piezoelectric charge constant is known to be the impression of intrinsic piezoelectric effect and extrinsic reversible domain wall motion (irreversible domain wall motion may also be considered if the signal is not small enough) and is called small signal piezoelectric co-efficient. When the input signal is much larger than the coercive field of the material, irreversible domain wall motion along with partial switching behavior results in nonlinear dependency of d on the amplitude of applied field (mechanical or electrical). Such a co-relation defines another useful parameter, that is, the large-signal piezoelectric coefficient d_{ij}^*. This quantity has a resemblance to S_{max}/E_{max}, which is one of the primary requirements for designing actuators with maximum strain at a minimum coaxial electric field. There is a large community of researchers working on the improvement of actuating performance of the lead-free piezoceramics.

It is a well-acclaimed fact that piezoelectrics have outstanding advantages in terms of sensor applications. Piezo-sensors are usually designed to detect stresses (σ_j) by monitoring the dielectric displacement (D_i) and, hence, can be represented by the relation:

$$D_i = d_{ij}\sigma_j \tag{5}$$

d_{ij} in equations (3) and (5) are mathematically equivalent and from a sensor point of view, d_{33} is often considered to be the most useful figure of merit. In most of the cases, d_{33} is also called as the longitudinal piezoelectric co-efficient, that is, when both the stress (strain) generated and the dielectric displacement (electric field) are along 3-direction. It is observed that the piezoelectric co-efficient has a pronounced dependence on the amplitude and frequency of electrical/mechanical driving field, bias electric field, temperature, pressure, and so forth. Sensors are the requirements of almost all smart systems, and one such application of sensory system in footwear is to monitor the speed, distance, lateral movement, acceleration, jump height, foot strike pattern, and loft time measurement during running, and so forth, for athletic performance and has been patented by Riley et al. [5] in 2010. Apart from the piezoelectric coefficient, the other material parameter for measuring the sensitivity of sensors and igniters is the so-called *piezoelectric voltage constant* 'g,' which is derived by dividing the dielectric permittivity ($\varepsilon_r\varepsilon_0$, where ε_r and ε_0 are the relative permittivity and permittivity of free space, respectively). The voltage constant g measures the maximum achievable voltage output for a given input mechanical signal and, alternatively, a mechanical displacement as a function of accumulated charge density. In addition, the product of charge constant and voltage constant ($d*g$) is another effective factor for a number of non-resonant transducer applications, such as microphones and energy harvesting.

2.2.2 ELECTROMECHANICAL COUPLING FACTOR (κ)

The square of the electromechanical coupling factor (k_{ij}^2) is a measure of energy stored in the piezoelectric material on the application of a driving field (electric field /mechanical displacement). k_{ij}^2 is related to d_{ij} and g_{ij} as:

$$k_{ij}^2 = \frac{d_{ij}^2}{\varepsilon_r\varepsilon_0 S_{ij}} = \frac{d_{ij}g_{ij}}{S_{ij}} \tag{6}$$

This shows that the coupling factor is one of the main criteria of energy conversion efficiency between electrical and mechanical energies. Since the piezoceramics are usually mechanical vibrators, the vibration frequencies result in resonance as a function of the sample geometry. Thus, for real-life applications, the value of k quoted is often understood as the maximum possible value at a given vibration mode (31 or 33 mode) and is normally measured by the so-called resonance–anti resonance method; k_{ij} is a very important figure of merit for transducer applications that operate at elevated frequencies or under resonance conditions (~10 kHz to ~10 MHz). Transducers are mostly used in ultrasonic imaging, underwater listening, sonars, non-destructive testing, high-frequency filters, surface acoustic devices, and many more. One of the major concerns is the heat loss during device fabrication as it brings in issues of self-heating of the material (both in resonance and off-resonance modes). Hence, the performance of the device may not be evaluated by merely knowing the k value. Another salient figure of merit, that is, the mechanical quality factor, becomes the deciding parameter to a great extent.

2.2.3 MECHANICAL QUALITY FACTOR (Q_M)

The quality factor can be conventionally defined as the resistance to the damping of the oscillators or resonators. It is closely related to the sharpness of electromechanical resonance spectrum and obtained by plotting motional admittance Y_m around resonance frequency ω_0 with respect to full width $(2\Delta\omega)$ at $Y_m / \sqrt{2}$ as:

$$Q_m = \frac{\omega_0}{2\Delta\omega} \tag{7}$$

The significant heat generation at the resonance frequency mostly comes from the mechanical loss, which is the inverse of Q_m. Hence, Q_m is a very useful factor in estimating the potential of devices used for high-power and high-frequency (ultrasonic range) applications. Such applications normally prefer high Q_m and k values to allow the least possible mechanical loss.

2.3   OTHER MATERIAL PARAMETERS OF INTEREST

2.3.1   PERMITTIVITY

Capacitors are the essential passive components that contribute to almost 30 percent volume of power converters. They cater to manifold functions, including voltage smoothing, pulse discharge, filtering, coupling-decoupling, dc blocking, and power conditioning. The quality of the capacitor in terms of its permittivity (capacitance) greatly affects the performance of the devices. One of the major concerns is the stability of permittivity (capacitance) over a broad temperature range along with low losses to maintain its charge and avoid self-heating. However, the current technology is facing challenges to develop a reliable, high temperature (>> 200 °C) and high capacitance ceramic material to be used in a plethora of applications where electronics is exposed to high temperatures, such as deep oil drilling, aviation (distributed control and sensing systems that are placed near the engine), space exploration, automobiles (anti-lock brake system on wheels), nuclear reactors, and so forth [6-11]. The minimum prerequisite conditions for a ceramic to be used as a capacitor at elevated temperatures are listed as: (i) high permittivity with low loss and high volumetric efficiency in the usable temperature range; (ii) high resistivity to reduce the amount of leakage current; and (iii) temperature insensitive dielectric permittivity within a variation of ±15 percent over a wide temperature and frequency range. Hence, the Electronic Industries Association (EIA) has classified these materials into class I, II, III and IV, which are characterized by a two-letter and one-figure code indicating the minimum working temperature, maximum change allowed in permittivity (capacitance) and the maximum working temperature [12]. Further, to analyze the workable range of the ceramic capacitors with a temperature-stable dielectric permittivity, one has to calculate the temperature co-efficient of capacitance (TCC) as:

$$TCC = \frac{\Delta C}{C_{base\ temp.}} = \frac{\left(C_T - C_{base\ temp.}\right)}{C_{base\ temp.}} = \frac{\Delta \varepsilon}{\varepsilon_{base\ temp.}} \tag{8}$$

where, C_T is the capacitance at any temperature. However, it is worthy to note that the base temperature is not specified. In most of the traditional capacitors like EIA–X7R (-55–125 °C), X8R (-55–150 °C), X9R (-55–200 °C), the base temperature is chosen as 25 °C, but for the ceramics to be used at high temperatures, the maximum temperature is much higher than 250 °C [13, 14]. This has prompted the researchers to select a higher temperature as the base temperature (150–250 °C). The materials having excellent temperature stability of dielectric properties with TCC were proposed to be potential candidates for high temperature capacitor applications withstanding harsh conditions.

2.3.2 POWER OUTPUT

Energy harvesting from mechanical vibration to fuel the self-powered sustainable devices in order to meet the global energy crisis has gained worldwide attention. The materials used for energy harvesting require higher values of d_{33} and g_{33}. It is found that lead-free energy harvesters possess a high output voltage of 10–30 V, piezoelectric charge constant and voltage constant of the order of 250–900 pC/N and 24.89 × 10^{-3} Vm/N respectively [15, 16]. Apart from these parameters, the energy-harvesting performance of the piezo-materials is mainly evaluated from the output power generated under mechanical stimulation. The output power is calculated as [17]:

$$P_L = \frac{1}{T} \int \frac{V_0(t)^2}{R_L} dt \qquad (9)$$

where $V_0(t)$ is the output voltage signal, R_L is the load resistance and T is the time period of application of the mechanical stimulation. The output voltage is closely related to the piezoelectric properties of the material and, in relation to that, the open circuit voltage is given by [18]:

$$V_{oc} = \int g_{ij}\epsilon(l)E_p dl \qquad (10)$$

where g_{ij} is the piezoelectric voltage constant, $\epsilon(l)$ is the external strain, and E_p is the Young's modulus of the system. The open-circuit voltage increases in proportion to the compressive strain. Similar to the output voltage, the generated charge (Q) also increases with the external mechanical stimulation as [18]:

$$Q = d_{ij}\sigma_{ij}A \qquad (11)$$

where, d_{ij} is the piezoelectric charge constant, σ_{ij} is the external stress and A is the area of the electrodes. Much of the work in this regard has progressed using lead free piezoelectrics in the form of bulk ceramics, thin films, fibers and so forth. Device functionality of the piezoelectric harvesters is also estimated in terms of off-resonance figure of merit using the relation [19, 20]:

$$FOM_{off} = \frac{d_{33}^2}{\varepsilon_r \varepsilon_0} = \frac{d_{33}g_{33}}{tan\delta} \qquad (12)$$

2.3.3 ENERGY STORAGE EFFICIENCY

Capacitors are also preferred as energy storage devices to store energy, especially
from clean and renewable energy resources such as solar, wind, water, and thermal
energies. Due to faster charge–discharge rates (order of nano second) and higher
power density (up to 10^8 W/kg), they satisfy the requirement of super high-
power electronics and systems [21]. The energy storage efficiency of the ceramic
capacitors is calculated by employing the polarization electric field (P-E) loops as:

$$\text{Recoverable energy storage density}\left(W_{rec}\right) = \int_{P_r}^{P_{max}} E dP \qquad (13)$$

$$\text{Energy storage efficiency}\left(\eta\right) = \frac{W_{rec}}{W} \times 100 = \frac{W_{rec}}{W_{rec} + W_{loss}} \times 100 \qquad (14)$$

$$\text{Energy storage density}\left(W\right) = \int_0^{P_{max}} E dP \qquad (15)$$

where P_{max}, P_r, P and E denote the maximum electric field induced polarization,
remnant polarization, polarization and applied electric field respectively. The
important factors for obtaining high energy density are high P_{max} and low P_r. W_{loss} is
calculated from the numerical integration of the closed area of the P-E loop.

2.4 ROAD MAP TO ACHIEVE OPTIMUM FIGURES OF MERIT

In our preceding discussion, we have known that piezoelectric materials possibly
can be scaled up on the basis of their figures of merit, which need to be optimum
for multifarious applications. In view of enhancing these figures of merit, the
researchers may chalk out various strategies among which construction of PZT
like phase boundaries and/or enhancing the relaxor property by compositional
engineering will form a major part of this chapter.

2.4.1 CONSTRUCTING MORPHOTROPIC PHASE BOUNDARIES

The unprecedented properties of piezoelectric materials are accompanied with the
appearance of a polarization changes that become significant along the regions of
phase transitions. The best illustrations are the divergence observed in dielectric
and piezoelectric data in temperature-, mechanical stress-, electric field – and
composition-driven phase transitions. The compositionally propelled phase
boundary between two different crystal structures is of much practical importance
since it is almost immune to any external influences like temperature, electric field
or stress and is often addressed as the so-called morphotropic phase boundary
(MPB). The word "morphotropic" originally involves the phase transitions due to
changes in composition. Usually, MPB occurs at some critical composition when

there is phase instability of one phase against the other. At this composition, the two phases are energetically similar but structurally different due to which the mechanical strain to preserve one phase against the other is relaxed (softened) [22]. The first ever MPB was reported for PZT sixty years ago, and the existence of MPB was found to be responsible for the unparalleled piezoelectric response of PZT and its allied systems. In case of PZTs, the MPB is formed between tetragonal and rhombohedral phases, either due to compositional variations or mechanical pressure [23]. Enhancement in its electromechanical properties is basically due to the monoclinic phase bridging the tetragonal and rhombohedral phase, which promotes a large number of possible polarization directions. The monoclinic phase [24] has polarization vector aligned in a direction intermediate to those of tetragonal (pseudocubic [001] direction) and rhombohedral (pseudocubic [111] direction) phases and can aptly change its direction on poling leading to large piezo-response [25]. MPB can also arise under pressure at low temperatures, leading to symmetry allowed polarizations. Ideally, MPB should be approximately temperature independent so that the material composition always remains close to MPB with variations in temperature. But if the MPB is dependent on temperature, then the properties would be maximal only at the temperature of phase transition (MPB) as in case of BZT-xBCT system or BT-based systems, and as the temperature is increased or decreased, desirable properties may not be achievable [26]. Figure 2.5(a) features a differentiation between temperature-independent and temperature-dependent phase boundaries. Since then, the investigation on piezoelectric solid solutions displaying MPB has been the prime interest of the present time to fulfill the twofold goal of finding lead-free alternatives with operational stability at higher temperatures to meet the stringent environmental regulations. A similar type of phase boundary is also observed between tetragonal and rhombohedral phases in many of the NBT-based solid solutions.

Ishibashi and Iwata [27] in 1998 used the phenomenological theory of Landau and Davonshire and formulated a typical model to explain the morphotropic phase boundary, employing Gibb's free energy as a function of polarization given by:

$$F(P) = F_0 + \frac{\alpha}{2\varepsilon_0}\left[P_x^2 + P_y^2 + P_z^2\right] + \frac{\beta_1}{4\varepsilon_o^2}\left[P_x^4 + P_y^4 + P_z^4\right]$$
$$+ \frac{\beta_2}{2\varepsilon_o^2}\left(P_x^2 P_y^2 + P_y^2 P_z^2 + P_z^2 P_x^2\right) \tag{16}$$

where F_o is the free energy in paraelectric phase. α is temperature dependent and is related to the inverse of the Curie constant 'a' as $\alpha = a(T - T_c)$, 'T' is the thermodynamic temperature and 'T_c' is the Curie temperature; β_1 and β_2 are the model parameters as a function of composition. The above equation for free energy can be written in simplified form as:

$$F = F_o + \Delta F \tag{17}$$

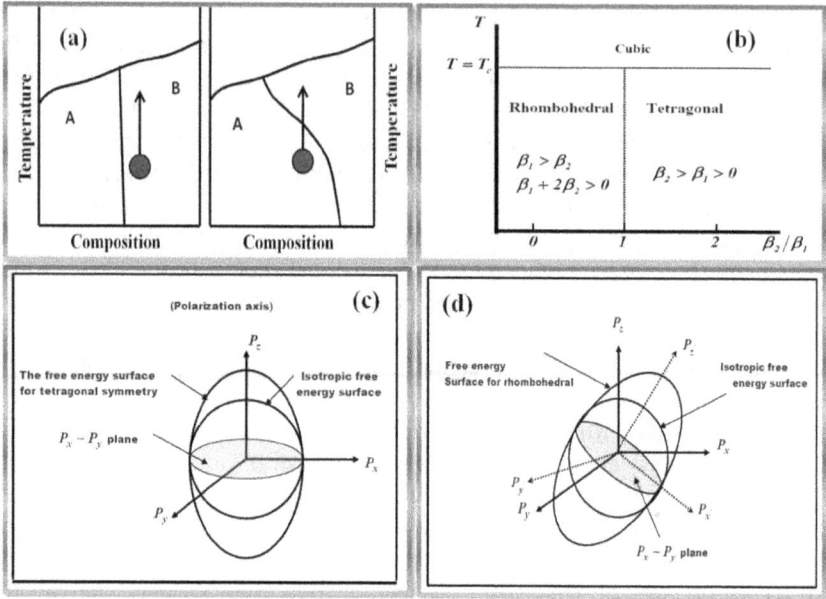

FIGURE 2.5 (a) Comparison between temperature independent and temperature dependent phase boundaries; (b) Proposed phase diagram in T-β_2/β_1 plane explaining morphotropic phase boundary as a function of material parameter $\beta^*=\beta_2/\beta_1$ (c) and (d) Orientation of free energy ellipsoid (for spontaneous polarization) with respect to isotropic energy surface ($\beta_1 = \beta_2$) for tetragonal and rhombohedral phases respectively.

Source: [27, 28, 29].

Considering a phase co-existence at $\beta_1 = \beta_2$ (in the free energy function) between tetragonal and rhombohedral phases, the static dielectric constant diverges at MPB. A phase diagram is proposed in the $T - \beta_2 / \beta_1$ plane explaining morphotropic phase boundary as a function of material parameter $\beta^* = \beta_2 / \beta_1$ is shown in Figure 2.5(b). The vertical dotted line in the figure at $\beta_1 = \beta_2$ shows the MPB where the tetragonal and rhombohedral phases co-exist and the energy surface is isotropic. The separation between the paraelectric (cubic) and ferroelectric phases is displayed by the solid line. Thermal stability of ferroelectric phase requires that for a particular direction of polarization, F→∞ as $|P| \to \infty$. Cubic-rhombohedral second order phase transitions occur in the β plane defined by $\beta_1 > \beta_2$ and $\beta_1 + 2\beta_2 > 0$ whereas cubic–tetragonal phase transitions occur in the $\beta_2 > \beta_1 > 0$ region [28]. To visualize the fact further, Ibrahim et al. [28] constructed a diagram taking P_x, P_y and P_z along the axes in orthogonal co-ordinate system and free energy is represented by a surface; whose shape is governed by the relative values of β_1 and β_2. The condition $\beta_1 = \beta_2$ (MPB) corresponds to isotropic energy surface and it appears as a sphere in the XYZ frame of reference. In the tetragonal phase, the spontaneous polarization, $P_{S.T}$, is along the (001) direction and hence, the free energy surface is elongated the Z-axis forming an ellipsoid. This tetragonal ellipsoid intersects with the isotropic sphere in the $P_x - P_y$ plane (Figure 2.5(c)). However, the spontaneous polarization $P_{S,R}$, for the rhombohedral phase is along the (111) direction, orienting the free energy ellipsoid along the same direction. The isotropic energy surface is also somewhat rotated with respect to the original frame of reference and is represented in the $P_x' - P_y'$ plane in Figure 2.5(d) [29]. Previously, there have been reports relating the material parameter $\beta^* = \beta_2 / \beta_1$ to MPB as a function of composition ion lead-based systems [27, 30, 31]. Theoretical studies on MPB in lead-free systems are very few but recently a research group from Japan reported a ferroelectric tetragonal phase with $P4bm$ symmetry, bridging the boundary between rhombohedral $R3c$ and tetragonal $P4mm$ phases in NBT-BT, unlike the monoclinic bridge in case of PZT [32].

2.4.2 Introducing Relaxor Behavior: Role of PNRs

Structural disorder within the materials due to the disruption of long-range systematic atomic arrangements because of the introduction of defects/imperfections has been the impetus for the charismatic properties of the solid systems since ages. While in amorphous/glassy materials, the structural coherence is lost beyond a few atomic neighbors; in crystals, the disorder appears as a result of perturbation in the periodic arrangement of atoms/ions [33]. Such turmoil in the local symmetry of crystals is basically responsible for functional response of technologically advanced materials. Relaxors form a major subgroup of such systems, whose properties generally argued to be governed by the nanometric cationic disorders giving rise to polar states (conventionally coined as polar nano regions) within the nonpolar matrix. Though there is no formal definition of these polar nano regions (PNRs), but they may be described as miniscule regions of finite size (static or dynamic) having non-zero spontaneous polarization (Figure 2.6). The presence of nanoscopic regions in relaxor materials is featured by broad frequency dispersion in the complex dielectric response with its maximum at T_m, slim hysteresis loop at T_m with small coercive fields, small remnant and spontaneous polarization, absence of spontaneous polarization in zero external electric field and logarithmic decay of polarization, which persists even above T_m [34]. A lot of theoretical and experimental work has been carried out by researchers to evaluate the origin and nature of the polar nano regions [35-41].

The origin of such peculiar behavior lies in the chemical inhomogeneity of the solid systems that arises due to charge and site disorder created due to substitution of cations in the A and B-sites with different valencies and sizes that break the translational symmetry and prevent the formation of long-range order [42, 43]. Some of the possible approaches may be to incorporate a dopant or a relaxor end member into a ferroelectric, forming a solid solution system. This fact is well evidenced in PMN systems (first relaxor ferroelectric discovered), which show dielectric relaxation over a wide range of temperatures; however, this behavior is enhanced when PT is added as an end member, along with significant improvement in the piezoelectric properties [44]. Similar response is also observed in some

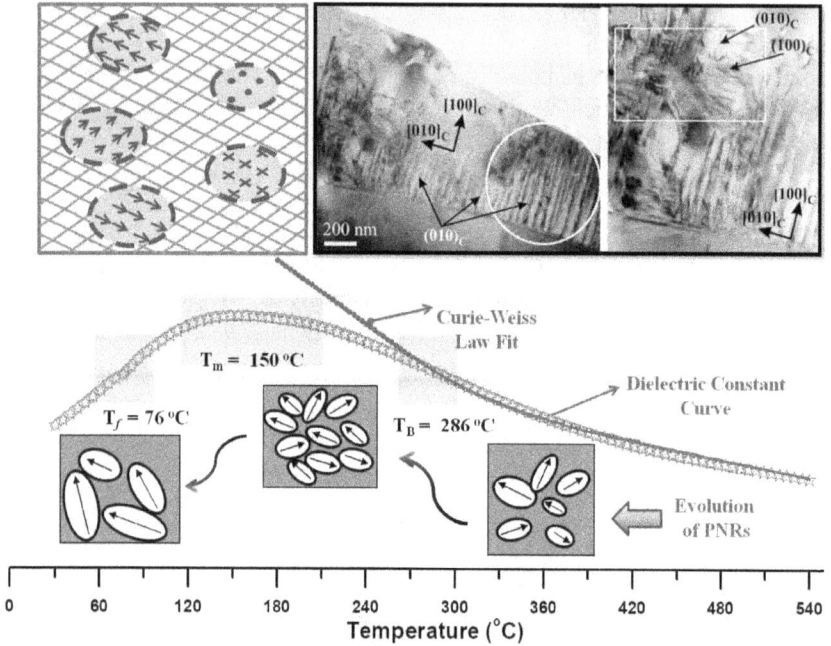

FIGURE 2.6 Schematic representation of formation of nanoscopic polar regions within the non-polar matrix (top left), TEM imaging of ferroelectric domains and distribution of nano-sized tetragonal platelets within rhombohedral matrix in $Na_{0.5}Bi_{0.5}TiO_3$ (top right), variation of dielectric constant with temperature for NBT-ST-0.08BT showing evolution of PNRs as a function of three characteristic temperatures T_B, T_m and T_f (T_B is obtained as a deviation of dielectric curve from the Curie Weiss law) (below).

Source: [53, 54, 55].

Pb-free systems like NBT, BZT, KNN etc. Detailed description about the relaxor behavior of NBT and NBT-based binary and ternary solid solutions are discussed in the following sections. A brief review about the role of charge disorder and random fields in the origin of relaxor state has been presented in some reports [45, 46]. At high temperatures, alike any normal ferroelectric material, the relaxors also exist in a non-polar paraelectric state. With decreasing temperature, some polar regions develop that are locally correlated and possess randomly distributed dipole moments. This transformation into an ergodic state occurs at a temperature so-called as the "Burns temperature, T_B" which is much above the temperature corresponding to maximum dielectric permittivity T_m [47]. But such a kind of development may not be accompanied by any kind of structural phase transition as it does not lead to any change in crystal structure on a mesoscopic or macroscopic scale. These PNRs are highly mobile and show ergodic behavior at temperatures close to T_B. As the temperature decreases, their dynamics slow down enormously and at temperatures much less than T_B, these PNRs freeze into a non-ergodic state, where the overall symmetry of the crystal still remains cubic. This phenomenon has been pointed out as a similarity between dipolar glasses and the canonical relaxors, where the thermally fluctuating dipole moments of the nano-sized clusters freeze out at very low temperatures [48]. This freezing process can be considered to be due to larger interactions between the nano regions with decreasing temperature and has been explained by the spin glass model [49-51]. The non-ergodic relaxor existing below the freezing temperature T_f can be transformed into a ferroelectric state by applying strong enough electric field. This property distinguishes relaxors from the typical dipolar glasses [46]. With increasing applied electric field, the volume of the PNRs (oriented in preferred direction at the cost of non preferred ones) increases. Such a type of volume change contributes to dielectric constant as well as polarization [52]. A typical figure showing the three characteristic temperatures T_B, T_m and T_f on the dielectric constant versus temperature curve along with schematic diagram of PNRs and their actual visualization by TEM is given in Figure 2.6 [53-55]. Relaxors are a material of choice for highly efficient multilayer capacitors, transducers, energy storage devices and electrostrictive actuators due to their outstanding dielectric and electromechanical properties. The high dielectric permittivity (ε) of these materials is maintained over a wide temperature span around T_m as compared to the classical ferroelectrics (highest dielectric permittivity is confined around T_c).

The microscopic origin of such a temperature independent permittivity may be attributed to the existence of PNRs and their thermally activated relaxation dynamics with a broad distribution of relaxation time. The temperature stability of ε is important from the point of view of multilayer capacitors, particularly in power electronics. Also, the nanoscopic size of the polar regions soothe their switching response to the applied electric field leading to minimal remnant polarization and slim P-E loops (not involve any kind of macroscopic domain switching). This makes the relaxor compounds extremely promising candidates for energy storage devices. They also offer a significant volume reduction (volume efficiency) in many capacitor applications, such as consumer electronics, pulsed power systems and energy harvesters [44]. Another aspect for evaluating the performance of these technologically advanced material systems is their notable electromechanical behavior, which is largely dependent on ε. Exemplifying the case of electrostrictive property, when the amplitude of the applied electric field is much higher than the coercive field of the material, electrostrictive strain dominates, and its dependence on permittivity can be studied by isolating the electrostrictive co-efficient Q_{ij} in 33 mode.

$$S_{33}\left(E_0\right)=Q_{33}P_3^2\left(E_0\right) \tag{18}$$

$$S_{33}\left(E_0\right)=Q_{33}\left[\int_0^{E_0}\varepsilon_{33}\left(E\right)dE\right]^2 \tag{19}$$

Where $S_{33}\left(E_0\right)$ is the longitudinal electric field induced electrostrictive strain; $\varepsilon_{33}\left(E\right)$ is the dielectric permittivity, $P_3\left(E_0\right)$ is the polarization under an electric field with amplitude E_0 and Q_{33} is the electrostrictive coefficient. Such a response has been observed in many of the Pb-based relaxors like PMN, PZN and PMN-PT single crystals/ceramics [56] and even in case of lead-free NBT-based solid solutions [57-59]. Again, it has been evidenced that large piezoelectric effect is obtained near the MPB composition as discussed in the above section. However, it is the unique feature of the relaxor ferroelectrics that the relaxor end members in the binary and ternary solid solutions infuse nanoscale heterogeneities, leading to additional energies (electrostatic and elastic energies associated with heterogeneous interfaces), which flatten the energy surface along with long range ferroelectric phase transition. Thus, the relaxor ferroelectrics may be interpreted as highly energetic systems with local fluctuations within the long-range ferroelectric matrix and it makes them very reactive to external stimuli like temperature, pressure, electric field and stress. Hence, the formation of those PNRs is one of the major ingredients responsible for the significant piezoelectric properties of the relaxors. Moreover, the relaxors have added advantages of reduced aging effect; do not require poling, and the properties are almost independent of grain size [44].

2.5   SUMMARY

Strain energy is one of the most ubiquitous ambient energies available to us, which can be captured and converted into useful power using piezoelectric technology. Hence, this article being the second chapter of the book is devoted to the fundamentals of piezoelectricity with reference to direct and indirect piezo-effects, and their splendid use in electro-ceramic industry. Further, the discussion continues on the piezoelectric figures of merit and their significance in energy-scavenging applications. To achieve optimum values of figures of merit specific for application, materials may be designed by compositional engineering. Above all, the current chapter may feed the readers with enough content that will be essential for understanding the piezoelectric figures of merit and develop novel materials for energy scavenging.

REFERENCES

1.   Kao, K.C., 2004. *Dielectric Phenomena in Solids*. Elsevier Academic Press, USA and UK.
2.   Lines, M.E. Glass, A.M., 1977. *Principles and Applications of Ferroelectric and Related Materials*, Oxford University Press, New York.
3.   Behera, R.K., Lee, C–W., Lee, D., Morozovska, A.N., Sinnott, S.B., Asthagiri, A., Gopalan, V., and Phillpot, S.R., 2011. Structure and energetics of 180° domain walls in PbTiO$_3$ by density functional theory, *Journal of Physics: Condensed Matter*, 23, pp. 175902.

4. Ma, C., Dulkin, E., and Roth, M., 2010. Domain structure-dielectric property relationship in lead-free $(1-x)(Bi_{1/2}Na_{1/2})TiO_3-xBaTiO_3$ ceramics, *Journal of Applied Physics*, 108, pp. 104105.

5. Riley, R.W., Hoffer, K.W., Berner Jr, W.E., Schrock, A.M., Niegowski, J.A. and Rauchholz, W.F., Nike, 2010. *Athletic performance sensing and/or tracking systems and methods*. U.S. Patent 7, 771, 320.

6. Hagler, P., Henson, P. and Johnson, R.W., 2010. Packaging technology for electronic applications in harsh high-temperature environments. *IEEE Transactions on Industrial Electronics*, *58*(7), pp. 2673–2682.

7. Buttay, C., Planson, D., Allard, B., Bergogne, D., Bevilacqua, P., Joubert, C., Lazar, M., Martin, C., Morel, H., Tournier, D. and Raynaud, C., 2011. State of the art of high temperature power electronics. *Materials Science and Engineering: B*, *176*(4), pp. 283–288.

8. Amalu, E.H., Ekere, N.N. and Bhatti, R.S., 2009, January. High temperature electronics: R&D challenges and trends in materials, packaging and interconnection technology. In *2009 2nd International Conference on Adaptive Science & Technology (ICAST)* (pp. 146–153). IEEE.

9. Johnson, R.W., Evans, J.L., Jacobsen, P., Thompson, J.R. and Christopher, M., 2004. The changing automotive environment: high-temperature electronics. *IEEE Transactions on Electronics Packaging Manufacturing*, *27*(3), pp. 164–176.

10. Chen, Y., Del Castillo, L., Aranki, N., Assad, C., Mazzola, M., Mojarradi, M. and Kolawa, E., 2008, April. Reliability assessment of high temperature electronics and packaging technologies for Venus mission. In *2008 IEEE International Reliability Physics Symposium* (pp. 641–642). IEEE.

11. Rittner, M. and Roth, A., 2010, March. Knowledge matrix for power electronics – The approach of the ZVEI working group 'High Temperature and Power Electronics'. In *2010 6th International Conference on Integrated Power Electronics Systems* (pp. 1–2). IEEE.

12. Jia, W., Hou, Y., Zheng, M., Xu, Y., Zhu, M., Yang, K., Cheng, H., Sun, S. and Xing, J., 2018. Advances in lead-free high-temperature dielectric materials for ceramic capacitor application. *IET Nanodielectrics*, *1*(1), pp. 3–16.

13. Muhammad, R., Iqbal, Y. and Reaney, I.M., 2016. $BaTiO_3–Bi(Mg_{2/3}Nb_{1/3})O_3$ ceramics for high-temperature capacitor applications. *Journal of the American Ceramic Society*, *99*(6), pp. 2089–2095.

14. Jia, W., Hou, Y., Zheng, M. and Zhu, M., 2017. High-temperature dielectrics based on $(1-x)(0.94Bi_{0.5}Na_{0.5}TiO_3-0.06BaTiO_3)-xNaNbO_3$ system. *Journal of Alloys and Compounds*, *724*, pp. 306–315.

15. Shin, D.J., Kang, W.S., Koh, J.H., Cho, K.H., Seo, C.E. and Lee, S.K., 2014. Comparative study between the pillar-and bulk-type multilayer structures for piezoelectric energy harvesters. *Physica Status Solidi (a)*, *211*(8), pp. 1812–1817.

16. Swallow, L.M., Luo, J.K., Siores, E., Patel, I. and Dodds, D., 2008. A piezoelectric fibre composite based energy harvesting device for potential wearable applications. *Smart Materials and Structures*, *17*(2), p. 025017.

17. Nunes-Pereira, J., Sencadas, V., Correia, V., Cardoso, V.F., Han, W., Rocha, J.G. and Lanceros-Méndez, S., 2015. Energy harvesting performance of $BaTiO_3$/poly (vinylidene fluoride–trifluoroethylene) spin coated nanocomposites. *Composites Part B: Engineering*, *72*, pp. 130–136.

18. Zhang, Y., Jeong, C.K., Wang, J., Sun, H., Li, F., Zhang, G., Chen, L.Q., Zhang, S., Chen, W. and Wang, Q., 2018. Flexible energy harvesting polymer composites based on biofibril-templated 3-dimensional interconnected piezoceramics. *Nano energy*, *50*, pp. 35–42.

19. Kandula, K.R., Asthana, S. and Raavi, S.S.K., 2018. Multifunctional Nd_{3+} substituted $Na_{0.5}Bi_{0.5}TiO_3$ as lead-free ceramics with enhanced luminescence, ferroelectric and energy harvesting properties. *RSC Advances*, *8*(28), pp. 15282–15289.

20. Kwon, Y.H., Lee, G.H. and Koh, J.H., 2015. Effects of sintering temperature on the piezoelectric properties of $(Bi,Na)TiO_3$-based composites for energy harvesting applications. *Ceramics International*, *41*, pp. S792-S797.

21. Zhang, Y., Liu, P., Shen, M., Li, W., Ma, W., Qin, Y., Zhang, H., Zhang, G., Wang, Q. and Jiang, S., 2019. High energy storage density of tetragonal PBLZST antiferroelectric ceramics with enhanced dielectric breakdown strength. *Ceramics International*, *46*(3), pp. 3921–3926.

22. Bhalla, A.S., Guo, R. and Alberta, E.F., 2002. Some comments on the morphotropic phase boundary and property diagrams in ferroelectric relaxor systems. *Materials Letters*, *54*(4), pp. 264–268.

23. Ahart, M., Somayazulu, M., Cohen, R.E., Ganesh, P., Dera, P., Mao, H.K., Hemley, R.J., Ren, Y., Liermann, P. and Wu, Z., 2008. Origin of morphotropic phase boundaries in ferroelectrics. *Nature*, *451*(7178), p. 545.

24. Heitmann, A.A. and Rossetti Jr, G.A., 2014. Thermodynamics of ferroelectric solid solutions with morphotropic phase boundaries. *Journal of the American Ceramic Society*, *97*(6), pp. 1661–1685.

25. Hu, Q., Alikin, D.O., Zelenovskiy, P.S., Ushakov, A.D., Chezganov, D.S., Bian, J., Zhao, Y., Tian, Y., Zhuang, Y., Li, J. and Jin, L., 2019. Phase distribution and corresponding piezoelectric responses in a morphotropic phase boundary Pb $(Mg_{1/3}Nb_{2/3})O_3$-$PbTiO_3$ single crystal revealed by confocal Raman spectroscopy and piezo-response force microscopy. *Journal of the European Ceramic Society*, *39*(14), pp. 4131–4138.

26. Rödel, J., Jo, W., Seifert, K.T., Anton, E.M., Granzow, T. and Damjanovic, D., 2009. Perspective on the development of lead-free piezoceramics. *Journal of the American Ceramic Society*, *92*(6), pp. 1153–1177.

27. Ishibashi, Y. and Iwata, M., 1998. Morphotropic phase boundary in solid solution systems of perovskite-type oxide ferroelectrics. *Japanese Journal of Applied Physics*, *37*(8B), p. L985.

28. Ibrahim, A.B.M., Murgan, R., Rahman, M.K.A. and Osman, J., 2011. Morphotropic phase boundary in ferroelectric materials. In *Ferroelectrics-Physical Effects*. IntechOpen.

29. Murgan, R., Tilley, D.R., Ishibashi, Y. and Osman, J., 2002. Evidence of Morphotropic Phase Boundary in the Linear Dielectric Susceptibilities of Bulk Ferroelectrics Materials. *Jurnal Fizik Malaysia*, *23*(1–4), pp. 166–170.

30. Ishibashi, Y. and Iwata, M., 1999. Phenomenology of Magnetostrictive Alloys of the Rare-Earth Fe 2 Compound Family. I. Elastic Constants. *Journal of the Physical Society of Japan*, *68*(4), pp. 1353–1356.

31. Ishibashi, Y. and Iwata, M., 1999. A theory of morphotropic phase boundary in solid-solution systems of perovskite-type oxide ferroelectrics. *Japanese Journal of Applied Physics*, *38*(2R), p. 800.

32. Kitanaka, Y., Miyayama, M. and Noguchi, Y., 2019. Ferrielectric-mediated morphotropic phase boundaries in Bi-based polar perovskites. *Scientific Reports*, *9*(1), p. 4087.

33. Eremenko, M., Krayzman, V., Bosak, A., Playford, H.Y., Chapman, K.W., Woicik, J.C., Ravel, B. and Levin, I., 2019. Local atomic order and hierarchical polar nanoregions in a classical relaxor ferroelectric. *Nature Communications*, 10(1), p. 2728.

34. Peláiz-Barranco, A., Calderón-Piñar, F., García-Zaldívar, O., González-Abreu, Y. and Peláiz-Barranco, A., 2012. Relaxor behaviour in ferroelectric ceramics. *Advances in Ferroelectrics*, pp. 85–107.

35. Guo, Z., Tai, R., Xu, H., Gao, C., Pan, G., Luo, H. and Namikawa, K., 2007. X-ray probe of the polar nanoregions in the relaxor ferroelectric $0.72Pb(Mg_{1/3}Nb_{2/3})O_3$–$0.28$ $PbTiO_3$. *Applied Physics Letters*, *91*(8), p. 081904.

36. Cheng, S.Y., Shieh, J., Lu, H.Y., Shen, C.Y., Tang, Y.C. and Ho, N.J., 2013. Structure analysis of bismuth sodium titanate-based A-site relaxor ferroelectrics by electron diffraction. *Journal of the European Ceramic Society*, *33*(11), pp. 2141–2153.

37. Dul'kin, E., Tiagunova, J., Mojaev, E. and Roth, M., 2017. Peculiar properties of phase transitions in $Na_{0.5}Bi_{0.5}TiO_3$–$0.06BaTiO_3$ lead-free relaxor ferroelectrics seen via acoustic emission. *Functional Materials Letters*, *10*(04), p. 1750048.

38. Macutkevic, J., Banys, J., Bussmann-Holder, A. and Bishop, A.R., 2011. Origin of polar nanoregions in relaxor ferroelectrics: Nonlinearity, discrete breather formation, and charge transfer. *Physical Review B*, *83*(18), p. 184301.

39. Prosandeev, S. and Bellaiche, L., 2016. Effects of atomic short-range order on properties of the $PbMg_{1/3}Nb_{2/3}O_3$ relaxor ferroelectric. *Physical Review B*, *94*(18), p. 180102.

40. Gui, H., Gu, B. and Zhang, X., 1995. Distribution of relaxation times in perovskite-type relaxor ferroelectrics. *Journal of Applied Physics*, *78*(3), pp. 1934–1939.

41. Gui, H., Zhang, X. and Gu, B., 1996. Dielectric response process in relaxor ferroelectrics. *Applied Physics Letters*, *69*(16), pp. 2353–2355.

42. Dai, X., DiGiovanni, A. and Viehland, D., 1993. Dielectric properties of tetragonal lanthanum modified lead zirconate titanate ceramics. *Journal of Applied Physics*, *74*(5), pp. 3399–3405.

43. Dai, X., Xu, Z. and Viehland, D., 1996. Normal to relaxor ferroelectric transformations in lanthanum-modified tetragonal-structured lead zirconate titanate ceramics. *Journal of Applied Physics*, *79*(2), pp. 1021–1026.

44. Li, F., Zhang, S., Damjanovic, D., Chen, L.Q. and Shrout, T.R., 2018. Local structural heterogeneity and electromechanical responses of ferroelectrics: learning from relaxor ferroelectrics. *Advanced Functional Materials*, *28*(37), p. 1801504.

45. Kleemann, W., 2006. The relaxor enigma – charge disorder and random fields in ferroelectrics. In *Frontiers of Ferroelectricity* (pp. 129–136). Springer, Boston.

46. Bokov, A.A. and Ye, Z.G., 2006. Recent progress in relaxor ferroelectrics with perovskite structure. *Journal of Materials Science*, *41*(1), pp. 31–52.

47. Burns, G. and Dacol, F.H., 1983. Glassy polarization behavior in ferroelectric compounds $Pb(Mg_{1/3}Nb_{2/3})O_3$ and $Pb(Zn_{1/3}Nb_{2/3})O_3$. *Solid State Communications*, *48*(10), pp. 853–856.

48. Viehland, D., Li, J.F., Jang, S.J., Cross, L.E. and Wuttig, M., 1991. Dipolar-glass model for lead magnesium niobate. *Physical Review B*, *43*(10), p. 8316.

49. Levstik, A., Kutnjak, Z., Filipič, C. and Pirc, R., 1998. Glassy freezing in relaxor ferroelectric lead magnesium niobate. *Physical Review B, 57*(18), p. 11204.

50. Kutnjak, Z., Filipič, C., Pirc, R., Levstik, A., Farhi, R. and El Marssi, M., 1999. Slow dynamics and ergodicity breaking in a lanthanum-modified lead zirconate titanate relaxor system. *Physical Review B, 59*(1), p. 294.

51. Bobnar, V., Kutnjak, Z., Pirc, R. and Levstik, A., 1999. Electric-field – temperature phase diagram of the relaxor ferroelectric lanthanum-modified lead zirconate titanate. *Physical Review B, 60*(9), p. 6420.

52. Xu, G., Wen, J., Stock, C. and Gehring, P.M., 2008. Phase instability induced by polar nanoregions in a relaxor ferroelectric system. *Nature Materials, 7*(7), p. 562.

53. Dorcet, V. and Trolliard, G., 2008. A transmission electron microscopy study of the A-site disordered perovskite $Na_{0.5}Bi_{0.5}TiO_3$. *Acta Materialia, 56*(8), pp. 1753–1761.

54. Praharaj, S., Rout, D., Subramanian, V. and Kang, S.J., 2016. Study of relaxor behavior in a lead-free $(Na_{0.5}Bi_{0.5})TiO_3$-$SrTiO_3$-$BaTiO_3$ ternary solid solution system. *Ceramics International, 42*(11), pp. 12663–12671.

55. Praharaj, S., Rout, D., Anwar, S. and Subramanian, V., 2017. Polar nano regions in lead free $(Na_{0.5}Bi_{0.5})TiO_3$-$SrTiO_3$-$BaTiO_3$ relaxors: An impedance spectroscopic study. *Journal of Alloys and Compounds, 706*, pp. 502–510.

56. Li, F., Jin, L., Xu, Z. and Zhang, S., 2014. Electrostrictive effect in ferroelectrics: An alternative approach to improve piezoelectricity. *Applied Physics Reviews, 1*(1), p. 011103.

57. Wang, F., Jin, C., Yao, Q. and Shi, W., 2013. Large electrostrictive effect in ternary $Bi_{0.5}Na_{0.5}TiO_3$-based solid solutions. *Journal of Applied Physics, 114*(2), p. 027004.

58. Jin, L., Pang, J., Luo, W., Lan, Y., Du, H., Yang, S., Li, F., Tian, Y., Wei, X., Xu, Z. and Guo, D., 2019. Phase transition behavior and high electrostrictive strains in $Bi(Li_{0.5}Nb_{0.5})O_3$-doped lead magnesium niobate-based solid solutions. *Journal of Alloys and Compounds, 806*, pp. 206–214.

59. Jin, L., Pang, J., Pu, Y., Xu, N., Tian, Y., Jing, R., Du, H., Wei, X., Xu, Z., Guo, D. and Xu, J., 2019. Thermally stable electrostrains and composition-dependent electrostrictive coefficient Q_{33} in lead-free ferroelectric ceramics. *Ceramics International, 45*(17), pp. 22854–22861.

3 Materials for Piezoelectric Energy Harvesting

3.1 CERAMICS

Piezoelectricity in materials is all about change in polarization on exposure to any mechanical deformation. This unique feature is specific to a class of active materials having non-centrosymmetric crystal structure. A general classification of piezoelectric energy harvesting materials is provided in Figure 3.1. However, to date the best known piezo materials are the ceramics owing to their sufficiently high dielectric and piezoelectric constants; and electromechanical coupling factors. Ceramics used in piezoelectric energy harvesting mostly possesses perovskite or wurzite structure. However, perovskites are more preferred due to their peculiar property of accommodating almost all elements in the periodic table. In the following paragraphs we will discuss the piezoelectricity in polycrystalline and single crystalline ceramics, basically perovskites.

3.1.1 POLYCRYSTALLINE

Piezoceramics with a polycrystalline structure are generally composed of several crystal grains having identical chemical compositions with different orientations of individual grains. Polycrystals may be both natural and synthetic. Despite the fact that quartz and berlinite exhibit piezoelectricity due to their unique crystalline structures, synthetic materials like $Pb(Zr_{1-x}Ti_x)O_3$ (PZT), and $BaTiO_3$ (BT) are still required to meet the need for materials for practical applications. Moreover, these synthetic materials have to undergo a poling process to impart a desired degree of polarization. $BaTiO_3$ was the first piezoelectric ceramic discovered in 1947 [1]. This was a revolutionary discovery since, for the first time, a polycrystalline ceramic was permanently rendered piezoelectric on poling. However, the Curie temperature of BT was quite low ($\sim$120 °C) which was not appropriate for high-temperature applications. Thereafter, a breakthrough was made when PZT demonstrating superior piezoelectric coefficients, dielectric permittivity, and factors as compared to $BaTiO_3$ was discovered in 1952. Near its morphotropic phase boundary ($x = 0.48$), where all of the aforementioned figures of merit are maximum, PZT is frequently utilized. It's coercive field and remnant polarization

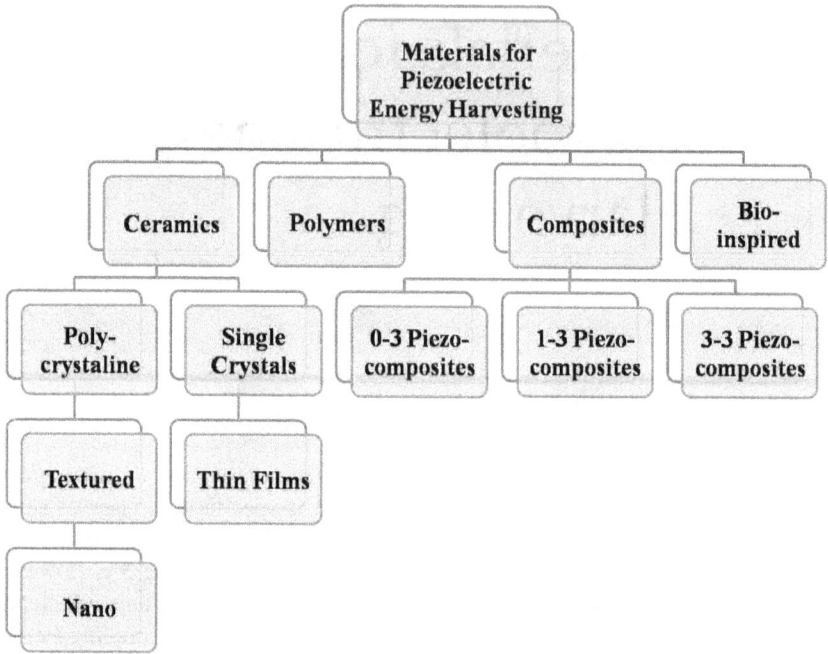

FIGURE 3.1 Schematic diagram showing general classification of piezoelectric energy harvesting materials.

are 35 C/cm^2 and 1 kV/mm, respectively [2]. Because of this, the poling field is still much below the dielectric breakdown strength, which causes a directed behavior during poling at high temperatures (piezo property). In some doped PZT materials, the piezoelectric coefficient for large signals can reach 779 pm/V [3]. Besides, doping with Pb(Ni$_{1/3}$Nb$_{2/3}$)O$_3$ resulted in a small signal piezoelectric coefficient d_{33} ranging from 710-1070 pC/N [4, 5]. Such features increased the demand for Pb-based systems, which lead to more and more release of Pb in the form of PbO during calcination and sintering. PbO in the environment is risky and can cause health issues. It is listed as one of the hazardous materials by not only the European Union (EU) but also by the Indian government and many other progressing countries. This acted as a trigger to search for environment-friendly, lead-free polycrystalline ceramics with superior properties to replace the PZT family. Though there are many good reports on the lead-free piezoceramics, they are application specific. Among the various known lead-free polycrystalline systems, the sodium potassium niobate (K$_{1-x}$Na$_x$NbO$_3$ or KNN) family and Bi based systems including Na$_{0.5}$Bi$_{0.5}$TiO$_3$ (NBT) and K$_{0.5}$Bi$_{0.5}$TiO$_3$ (KBT) are prospective. KNN polycrystals exhibit a dielectric constant of ~230–475, a coupling coefficient of ~23–40 percent and small signal piezoelectric coefficient of ~80–160 pC/N [6–9]. However, these properties were inferior to those of PZT and, not only that, it was also difficult to process KNN ceramics. Hence, chemical modification was one of the effective strategies to improve their performance. In this regard, Saito et al. [10] restructured polycrystalline KNN with Li, Ta, and Sb and obtained enhanced d_{33} values, i.e. >300 pC/N for untextured samples and >400 pC/N for textured samples with a Curie temperature (T_C) of 253 °C around MPB between orthorhombic and tetragonal phases. Another important figure of merit, i.e. the quality factor Q_m can also be improved by doping with oxides such as CuO. CuO in KNN exhibits a pining effect on the ferroelectric domains inducing double/pinched P-E loop having an extraordinarily large Q_m of 2235 though the value of d_{33} was very less [11]. Apart from that, other polycrystalline compounds such as NBT and KBT are proved to be key end members to design a variety of compositions. Though they exhibit good dielectric and piezoelectric properties, the existence of a high coercive field makes them difficult to be poled. To achieve properties at par with lead-based counterparts and diminish the coercive field, both compounds are often chemically modified by forming binary or ternary solid solutions. In this connection, Kanuru et al. [12] synthesized (1-x)NBT-xBT around the MPB composition ($x = 0.06, 0.07$ wt%) by solid state reaction and observed that pebble-like morphology changed to regular cubic structures as the composition was changed from $x = 0.06$ to 0.07. An improvement in piezoelectric properties was also evidenced from P-E loops with high remnant polarization and improvement in d_{33} from 85 pC/N ($x = 0.06$) to 102 pC/N ($x = 0.07$) (Figure 3.2). Ternary ceramic polycrystals are more attractive than that of binary systems. To cite a few examples, 0.9NBT-0.05KBT-0.05BT ceramics prepared by sol-gel flame method deliver a remnant polarization of 23.55 µC/cm² and $d_{33} = 213$ pC/N [13]. Besides, large electric field induced strain was obtained in some NBT-based ternary systems, for e.g. 688 pm/V for 0.76NBT-0.2ST-0.04BT [14] and 858 pm/V for 0.86NBT-0.07BT-0.07BFO [15] respectively.

3.1.2   SINGLE CRYSTAL

Single crystals are materials with continuous and uninterrupted crystal lattice throughout the sample with no grain boundaries. Piezoelectric single crystals (directionally oriented) display exceptional dielectric and piezoelectric properties in comparison to their bulk counterparts owing to the lack of grain boundary defects and the existence of inherent anisotropy in the materials. Many research works have been carried out by different groups in the last four decades. Shrout et al. [16] investigated piezoelectric PMN-PT and PZN-PT single crystals and obtained a piezoelectric coefficient higher than 2000 pC/N; electromechanical coupling factor >90 percent and a strain reaching more than 1 percent. These quoted values of figures of merit outperformed the PZT-based polycrystalline ceramics. A new generation of materials for electromechanical transducers was created as a result of these developments. Even though PMN-PT and PZN-PT single crystals have exceptional piezoelectric capabilities, they are limited by low Curie and phase transition temperatures, a weaker coercive field, and a lower quality factor. The lower value of phase transition temperatures makes the crystal prone to depoling

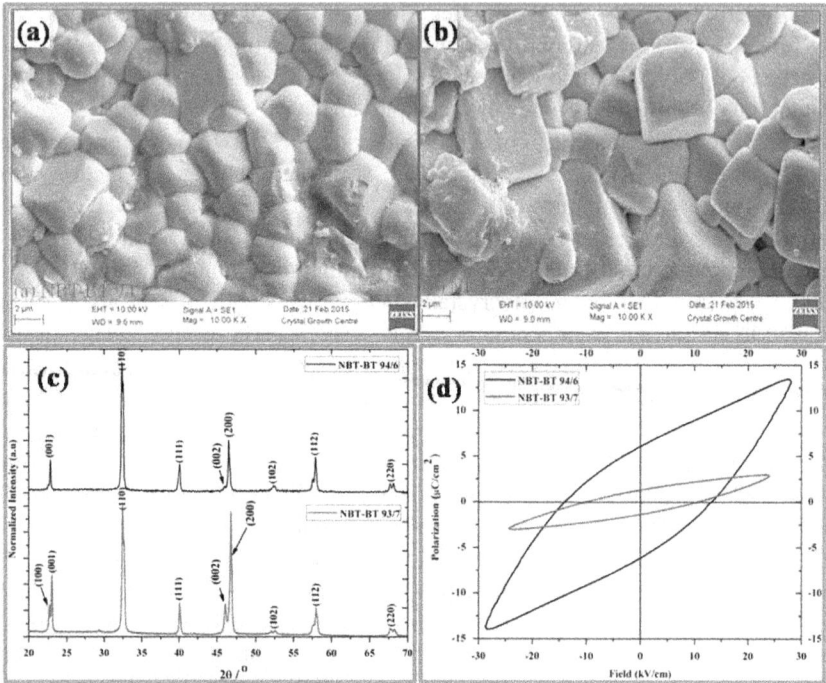

FIGURE 3.2 Morphology change of polycrystalline ceramics with change in composition (a) NBT-BT 94/6; (b) NBT-BT 93/7; (c) X-ray diffractograms and (d) P-E hysteresis loops of NBT-BT 94/6 and 93/7 respectively.

Source: [12].

under moderate temperature conditions and limits the working temperature range. Hence, the application of an extra dc electric field is necessary to sustain the polarization, which increases the complexity of the process and incurs extra cost. Therefore, single crystals with temperature and polarization stability are highly essential. In this connection, ternary relaxor crystals such as $Pb(In_{1/2}Nb_{1/2})O_3$-$Pb(Mg_{1/3}Nb_{1/3})O_3$-$PbTiO_3$(PIN-PMN-PT) and Mn doped PMN-PZT demonstrate higher Curie temperature, larger coercive fields and a superior quality factor [17, 18]. Very recently, Tian et al. [19] studied the piezoelectric, frequency, and impedance behavior of PMN-PT single crystals at different pressure conditions to understand their properties further. Though both the materials in the form of a ring, they exhibited good piezoelectric properties in free state. The single crystal ring was found to have lower resonant frequency and higher d_{33} (almost 4 times) as compared to PZT-4 ceramic rings. However, under static pressure conditions, the properties of the single crystal were not as stable as the polycrystalline ceramic. In spite of great success of the Pb-based crystals, there is an urgent need to find lead-free alternatives to combat the environment-related issues. The most common

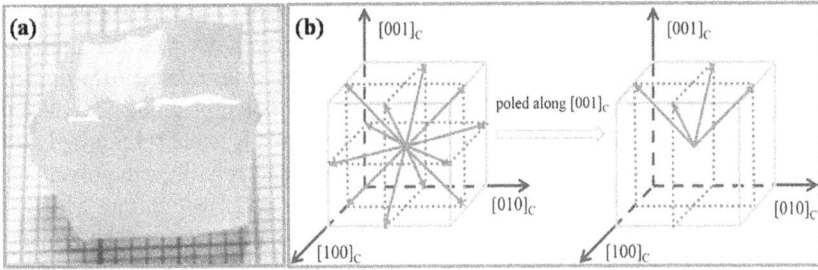

FIGURE 3.3   (a) Picture showing as grown Ta doped KNN single crystal; schematic diagram displaying spontaneous polarization directions of perovskite orthorhombic phase (a) before poling; (b) after poling.

Source: [21].

lead-free crystal that could be researched is $BaTiO_3$. BT single crystals have been investigated for more than 50 years and it showcases a piezoelectric coefficient in the range 68.5–316.6 pm/V. In 2009, Tazaki et al. [20] examined the variation of longitudinal as well as transverse piezoelectric effects with lattice distortion of the BT monodomain. They detected lattice distortions in presence of an applied electric field using a synchrotron X-ray. Lattice distortions appear in the form of lattice constant variation with a alterations in the applied field that could be co-related with the linear response of piezoelectric constant. On application of electric field parallel to the direction of spontaneous polarization, the c-axis stretches while the a-axis shortens at a rate of $d_{31} = -82 \pm 61\, pm/V$ and $d_{33} = 149 \pm 54$ pm/V. This study provided an insight into the intrinsic piezoelectric response on lattice scale for single crystals. Going forward in this direction, environmentally benign large single crystals ($12 \times 11 \times 11$ mm^3) of Ta doped KNN were synthesized by top-seeded solution growth technique. The large size of the crystal was advantageous in employing domain engineering with the intention of enhancing the piezoelectric, dielectric and elastic properties. The crystal poled along $[001]_c$ exhibited superior electromechanical coupling factors ($k_t = 0.646$, $k_{33} = 0.827$) and tan δ as low as 0.004. A single crystal of Ta doped KNN along with the poling direction are shown in Figure 3.3. These properties were highly prospective from the point of view of electromechanical devices [21]. Similar investigations on potential lead-free piezoelectric single crystals, including NBT-KBT, NBT-BT, NBT-BT-KNN systems [22-25].

3.1.3 TEXTURED

So far, we have learned that perovskite single crystals possess much superior piezoelectric properties than the polycrystals due to the highly anisotropic nature of the single granular structure. However, the piezoelectric behavior in case of polycrystalline ceramics is the average over all the grains. Besides, single crystals involve a high cost of production and complex synthesis process, which limit their extensive usage in piezoelectric devices. To address the issues related to the high cost of single crystals and low piezoelectric properties of polycrystalline ceramics, the fabrication of textured ceramics may be one of the most feasible solutions. Texture in ceramics refers to the orientation distribution of the crystallites in polycrystals. Textured ceramics are usually grown by templated grain growth during which tiny crystallites are oriented in a particle matrix. On heating the mixture, the growth of crystals throughout the mixture is guided in a particular direction by the particles. This results in a structure at par in uniformity of a single crystal. Hence, to explore the unique properties of textured ceramics, Yang et al. [26] employed a template growth method to fabricate <001> textured PIN-Pb($Sc_{1/2}Nb_{1/2}$)O_3-PT ceramics containing a huge amount of Sc_2O_3 which is a refractory compound. The presence of refractory material raises the phase transition temperature of rhombohedral to tetragonal structure (around 160–200 °C) which is beneficial from the point of view of wide working range of temperature. Further, the system exhibits a high k_{33} = 85-89 percent which makes it a suitable alternative to PT single crystals. Looking forward, textured lead-free ceramics have also attracted the attention of many. In this regard, ultrahigh piezoelectric coefficients (d_{33} = 700 pC/N, d_{33} = 980 pm/V and k_p = 76%) with superior temperature stability and electrical reliability were observed in KNN-based textured ceramics. Such outstanding features may be devoted to strong anisotropy, optimized engineered domain configuration and preferred polarization rotation facilitated by an intermediate phase. Apart from that, the reduced domain wall energy of the nanodomains along with increased domain wall mobility also play a major role in achieving high performance textured piezoelectrics [27]. Similar works were also performed in NBT and KBT-based systems [28, 29]. The improvements observed in relative permittivity, loss tangent, and d_{33} for the textured ceramics as compared to the non-textured ones are shown in Figure 3.4 [29].

3.1.4 Nano

In recent years, fusion of nanotechnology with piezoelectric technology has led to a new class of materials called "piezoelectric nanomaterials." This novel category of materials demonstrates superior piezoelectric properties along with splendid resilience and peculiar coupling between semiconducting and piezoelectric properties. Under an externally applied force, a potential is generated in piezo nanomaterials due to distortion of the crystal lattice. Basing on the dimension of the nanomaterials, they can be subdivided into zero dimensional (0D), which include nanoclusters, nanodispersions and so forth; one-dimensional (1D) covering nanorods and nanowires; two-dimensional (2D) including nanoribbons and films; and finally three-dimensional (3D) nanostructures (Figure 3.5). All these dimensionally diverse nanomaterials contribute differently to piezoelectricity. Most of the reports on nanomaterials employed for piezoelectric applications

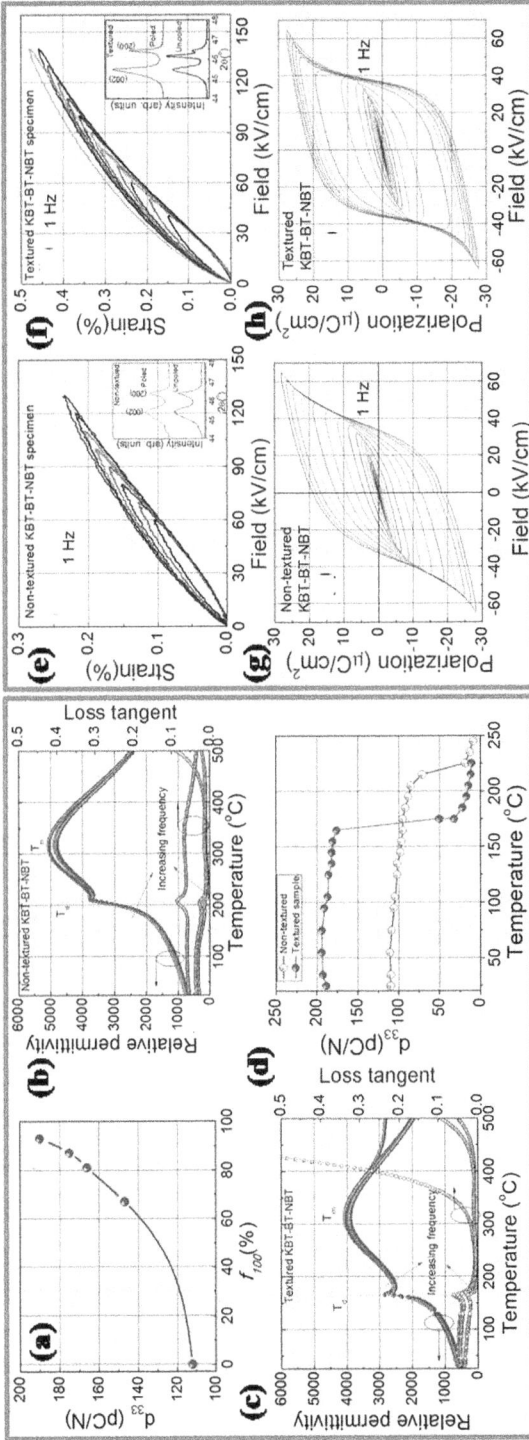

FIGURE 3.4 (a) d_{33} as a function of degree of texturing for KBT-BT-NBT ceramics; temperature variation of relative permittivity and loss tangent at 1, 10 and 100 kHz for (b) non-textured and (c) textured KBT-BT-NBT; (d) variation of d_{33} with temperature for textured and non-textured KBT-BT-NBT; (e) and (f) comparison among unipolar S-E loops for non-textured and textured KBT-BT-NBT ceramics respectively; (g) and (h) P-E loops for non-textured and textured KBT-BT-NBT respectively.

Source: [29].

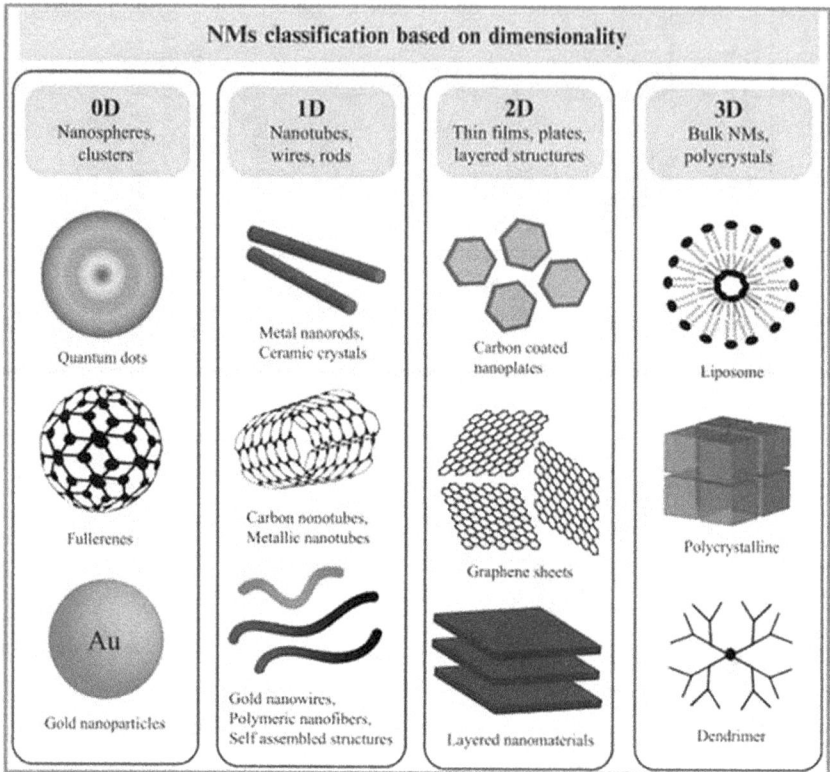

FIGURE 3.5 Classification of nanomaterials on the basis of their dimensions.

focus on ZnO since it is relatively easier to synthesize those nanostructures with crystallographic orientations at low temperatures. The initial studies on ZnO nanoachitectures i.e. aligned nanowire arrays grown on Al_2O_3 substrates, were carried out by Wang and Song in 2006 [30]. The nanowires were synthesized using a vapor-liquid-solid based process using an Au catalyst. On performing piezoelectric measurements by atomic force microscopy (AFM) using Silicon (Si) tip coated with platinum (Pt), it is found that most of the Au particles either evaporated from the nanowire tips or deflected and dropped off by the AFM tip. The AFM tip could reach a single nanowire without interference due to the less densely packed nanowires and relatively lower lengths ~0.2–0.5 µm. The voltage generated by these materials was mainly due to the generation of a strain field and detachment of charge attributed to bending by the AFM tip. The Schottky barrier formed between ZnO nanowires and AFM tip is also another major contributor of piezoelectric voltage. Similarly, 2D nanosheets of ZnO have been reported by Kim et al. [31] in which the authors sandwiched ZnO nanosheet/anionic nanoclay layer between Au-coated PES and Al-coated PES as top and bottom electrodes, respectively, to construct piezoelectric nanogenerator (PENG) (Figure 3.6). When a compressive force of 4 kgf was applied, this PENG produced 0.7 V and 17 A cm^{-2} in voltage and current, respectively. In a similar fashion, electromechanical properties of ZnO nanorods and ZnO nanowalls grown on ITO and aluminum substrate respectively were compared by Fortunato et al. [32]. Though both samples were prepared by chemical bath deposition, they exhibited different piezoelectric constants for ZnO nanorods (7.01 ± 0.33 pC/N) and nanowalls (2.63 ± 0.49 pC/N). Better piezo features of the nanorod structure were due to their superior orientation along the c-axis and lesser defect rate than the nanowalls. It can be inferred that 1D nanostructures deliver better piezoelectric figures of merit than 2D structures. Among the other piezoelectric nanomaterials, electrospun PZT is known for its high d_{33} values, flexibility and mechanical strength. Chen et al. [33] fabricated a PENG using PZT fibers of length 500 μm and diameter 60 nm, respectively, enclosed in a soft PDMS matrix. The PENG delivered 1.63 V output voltage and power of 6 MΩ. In the league of lead-free materials, barium

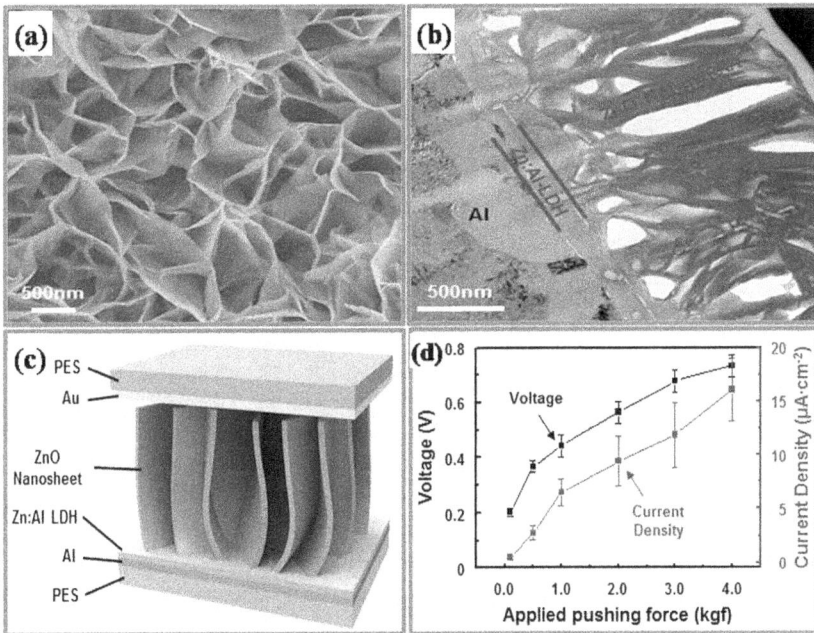

FIGURE 3.6   (a) Planer FESEM image of ZnO nanosheets grown on Al electrode; (b) TEM micrograph of the cross section of ZnO nanosheet network with the formation of LDH at the interface of ZnO and Al electrode; (c) schematic diagram showing nanogenerator constructed from 2D ZnO nanosheet network /anionic nanoclay heterojunctions; (d) output voltage and current density generated by the nanogenerator on application of a pushing force.

Source: [31].

titanate synthesized by hydrothermal process demonstrated large d_{33} of 460 pC/N and d_{31} of −185 pC/N with an electromechanical coupling factor k_p of 42 percent. In addition, the poled samples possessed a dielectric constant as high as 5000 and a Poisson's ratio of 0.38, which might be the origin of superior figures of merit [34].

3.1.5 THIN FILMS

Piezoelectric thin films are a class of materials advantageous for microscale and nanoscale devices, particularly MEMS (micromechanical systems) and NEMS (nano electromechanical systems). Earlier, PZT thin films have been very popular in devices owing to high piezoelectric coefficients. PZT based thin films offer advantages in terms of good actuating range, less power consumption, high blocking force and relatively wider operating frequency range. In 2014, Lee and his coworkers [35] used the laser lift-off (LLO) technique to produce large area PZT thin films on flexible substrates. The film was lightweight, flexible, and highly efficient to be used as an energy-harvesting device. The output voltage and current were measured to be 200 V and 150 A/cm² throughout its recurring bending and unbending operations. Such output features demonstrate superior piezoelectric performance required for an energy harvesting application. Further, the nanogenerator device constructed out of this thin film could power a series of 105 commercial-LED arrays around 250 V without rectifier and charge circuits. To combat the disastrous side effects of the PZT family, the recent research is focused on lead-free materials. Barium titanate being an excellent lead-free piezoelectric, both in bulk and nano forms, is also widely studied from the point of view of thin films. In this regard, Cernea et al. [36] fabricated $0.89(Na_{0.5}Bi_{0.5})TiO_3\text{-}0.11BaTiO_3$ thin film with a surface roughness of 5 nm by spin coating process on a Si wafer coated with platinum. The effective piezoelectric coefficient d_{33} noted at 5 V of dc voltage was found to be around 29 pm/V. The low value of d_{33} is related to the high leakage current (5.2×10^{-4} A/cm²), which also degrades the dielectric and ferroelectric properties of the film. On the other hand, using $Pt/Ti/SiO_2/Si$ substrates, homogeneous $(Na_{0.5}K_{0.5})NbO_3$ films are grown by RF sputtering and demonstrate an increase in d_{33} from 74 ± 11 to 120 ± 18 pm/V when transferred on to a polyamide substrate. This is associated with a change in structure from pseudocubic to orthorhombic. Apart from that, the transferred film exhibits a very low leakage current of 1.1×10^{-8} A/cm² and reduced coercive field and remnant polarization. The energy harvester constructed using this film delivered an optimum output voltage of 1.9 V and 38 nA current corresponding to a power density of 2.89 µW/cm³ [37]. Qin et al. [38] deposited large area ZnO thin films on a Si wafer at 500 °C substrate temperature by pulsed laser deposition. The films deposited display preferred c-axis orientation and a high piezoelectric coefficient of 49.7 pm/V (Figure 3.7). However, still the research in this direction is in its nascent state and needs much more progress.

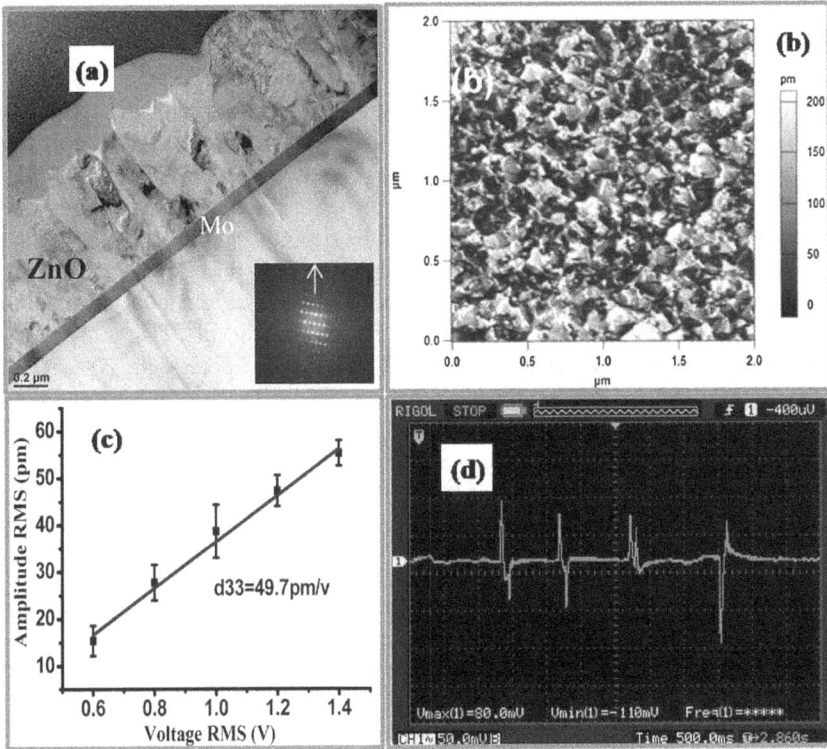

FIGURE 3.7 (a) TEM image of ZnO thin film cross section grown on Mo substrate; (b) piezoelectric response of ZnO thin films characterized by PFM; (c) linear fit of PFM amplitude vs applied AC voltage and calculated d_{33}; (d) open circuit voltage obtained from the ZnO thin film recorded by oscilloscope.

Source: [38].

3.2 POLYMERS

Some carbon-based polymers display a piezoelectric effect as a result of their molecular arrangement and orientation. They exhibit moderate strain coefficients (d_{33}) and voltage coefficients (g_{33}) and are significantly softer than ceramics. However, due to their special qualities, including design versatility, low density, and ease of processing, they are suitable for a variety of energy harvesting applications. Piezoelectric polymers may be classified into semi-crystalline and amorphous polymers. The majority of the polymers showing piezoelectricity are semi-crystalline in nature having a polar crystalline phase. Mechanical orientation, high voltage treatment or thermal annealing are highly beneficial in inducing strong polar phase into the polymers. On the other hand, amorphous polymers exhibit reduced piezo response due to their molecular arrangement. Besides, the two basic criteria for any polymer to qualify as a piezoelectric are (i) existence of inherent molecular dipoles within the polymer structure, and (ii) preferred orientation of the dipoles to maintain a particular orientation state. Such an orientation can be achieved by the application of an external dc bias (poling). Research in the direction of piezoelectric polymers gained importance after the pioneering work of Kawai and his coworkers [39] in the late 1960s. They have fabricated a semi-crystalline thermoplastic polymer from vinyldene difluoride (VDF) through the polymerization. Moreover, PVDF retains 50–70 percent crystallinity existing in five polymorphic phases: α, β, γ, δ and ε. The α-phase is generally non-polar in nature due to the alternate arrangement of H_2 and F_2 atoms on both the sides of the polymer chain. (TGTG′ trans-gauche configuration). On the other hand, β, γ and δ phases are polar with zigzag TTT, TTTGTTTG′ and TGTG′ conformations respectively. The β phase possesses the maximum dipole moment per unit cell i.e., 8×10^{-30} C m. A schematic diagram representing α, β and γ configurations are displayed in Figure 3.8. Therefore, efforts are made to increase the amount of β phase in order to improve the piezoelectric

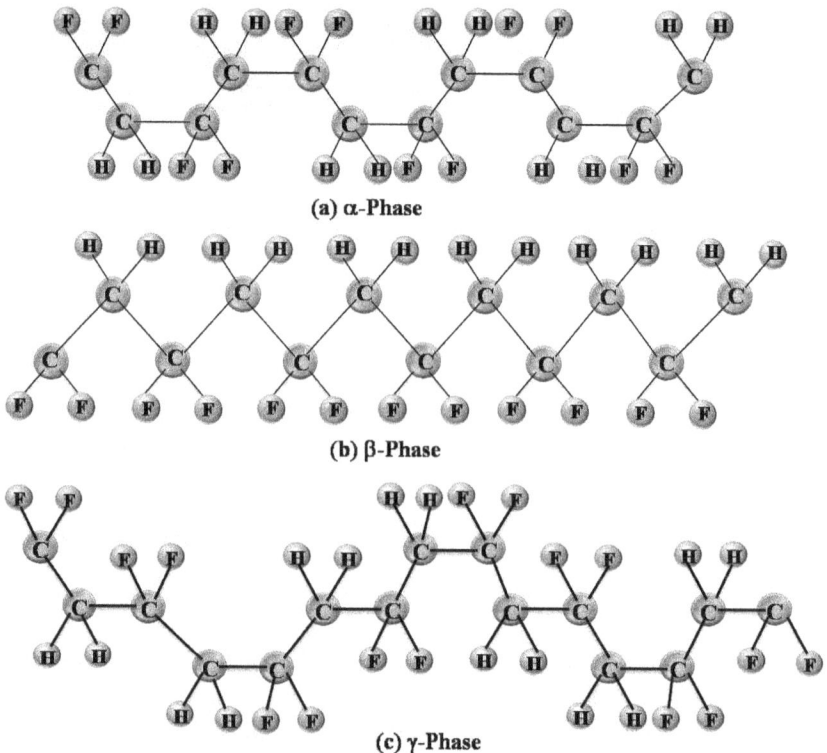

(a) α-Phase

(b) β-Phase

(c) γ-Phase

FIGURE 3.8   Schematic diagram displaying α, β and γ phase of PVDF.

performance of PVDF. A maximum d_{31} of 60 pC/N under a poling field of 0.55 MV/ cm was reported at a temperature of 80 °C and stretching ratio of 4.5 by Kaura et al. [40] in 1991. However, the highest amount of this phase was obtained by Gomes et al. [41] with a stretching ratio of 5 and d_{33} value of 34 pC/N. In another work, a better piezoelectric response was obtained by solution casted PVDF films in a poled state as compared to its unpoled state. The measured d_{33} value was –5 pC/N at a poling field of 20 kV/cm [42]. P(VDF-TrFE), a semicrystalline co-polymer of PVDF exhibits better piezoelectric property ($d_{33} = -30$ to–40 pC/N and $d_{31} = 25$ pC/ N) by transforming the α-phase into β-phase. By soaking a template into 200 nm-wide channels of anodic porous alumina membranes, Cauda et al. [43] created PVDF and P(VDF-TrFE) nanowires in a 70/30 ratio. The nanofinement caused the phase to crystallize in both polymers, improving their properties. Incorporation of hexafluoropropylene (HFP) into PVDF forms another co-polymer P(VDF-HFP) with properties dependent on the HFP content. Similarly other piezoelectric polymers include polylactic acid (PLA), polyurea, polyurethanes, polyamides and so forth. Various works have been undertaken by different researchers to improve the properties of these polymers. Polarity in PLA is induced by carbonyl groups

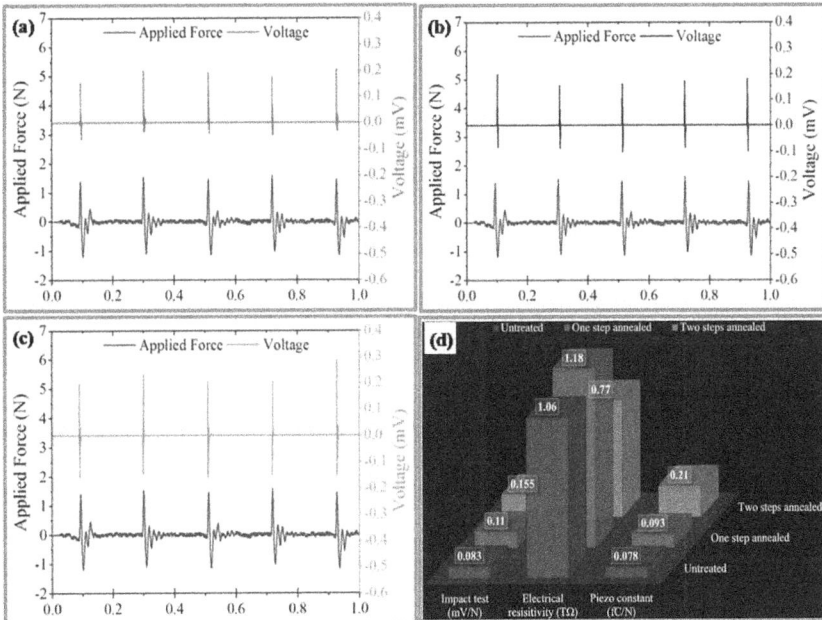

FIGURE 3.9 Impact test results of (a) PLA mat; (b) one step annealed PLA mat; (c) two step annealed PLA; and (d) schematic figure of impact test outcome, electrical resistivity and d_{33}.

Source: [44].

and a d_{14} of 10 pC/N is observed even in the absence of poling. But polyurea exhibits piezoelectric and pyroelectricity on poling. A class of semi-crystalline polymers with low piezoelectricity is polyamides and nylon. Though the research reports on the investigation of piezoelectric properties of polymers is sparse, but it is gaining importance day by day. Very recently, Farahani et al. [44] calculated d_{33} of commercial PLA and made an attempt to improve its piezoelectric properties by thermo-mechanical approach above the cold crystallization temperature. Such a treatment results in structural evolution from amorphous to crystalline states which, in turn, improves the piezoelectric property. The piezoelectric outcomes under different test conditions are displayed in Figure 3.9. Similarly, Yanase et al. [45] investigated upon the polyurea films synthesized by alternating deposition followed by thermal treatment. The measured piezoelectricity of the said samples was confirmed through direct measurement of transient current instigated by mechanical stress.

3.3   COMPOSITES

Piezoelectric composites mainly consist of a filler material with high piezoelectric constant embedded in a polymer matrix. Such an amalgamation is helpful in exploiting the advantages of both the materials. Combination of superior piezoelectricity and coupling factor of the filler materials along with polymer matrix flexibility makes the composites technologically important. The essential factors contributing to the exotic properties of the composites are shape, size, morphology and type of the fillers; and interaction of the polymer/filler interfaces. Based on the nature of the fillers, they may be grouped into three classes – conducting, non-conducting and hybrid. Conducting fillers include multiwalled carbon nanotubes, graphene, carbon black and so forth, while non-conducting fillers are $BaTiO_3$, ZnO, $(Pb,Zr)TiO_3$, $K_{0.5}Na_{0.5}NbO_3$ and so forth. On the contrary, hybrid fillers encompass more than one ingredient at nanometer/micrometer scale. Piezoelectric composites can also be classified on the basis of their dimensional connectivity as shown in Figure 3.10. Connectivity between the matrix and the filler is one of the important aspects affecting the performance of the composites. The piezoelectric phase of a composite with m-n connectivity is coupled in the m dimension for the polymer phase and in the n dimension for the filler particles. For instance, 0–3 connectivity refers to dispersion of piezoelectric particles in the polymer matrix while 1–3 connectivity indicates square/cylindrical/pillars of ceramic crystals implanted in polymer matrix. Synthesis of 0–3 composites is quite simpler than 1–3 configurations and can be done just by the dispersion of particles in the matrix. On the other hand, 1–3 composites require sophisticated methods like laminate-and-cut and dice-and-fill. These are the two most typical piezoelectric ceramic modes.

3.3.1   0-3 Piezo Composites

As per the classification suggested by Newnham et al. [46] and Pilgrim et al. [47], 0–3 composites constitute randomly scattered piezo components within a

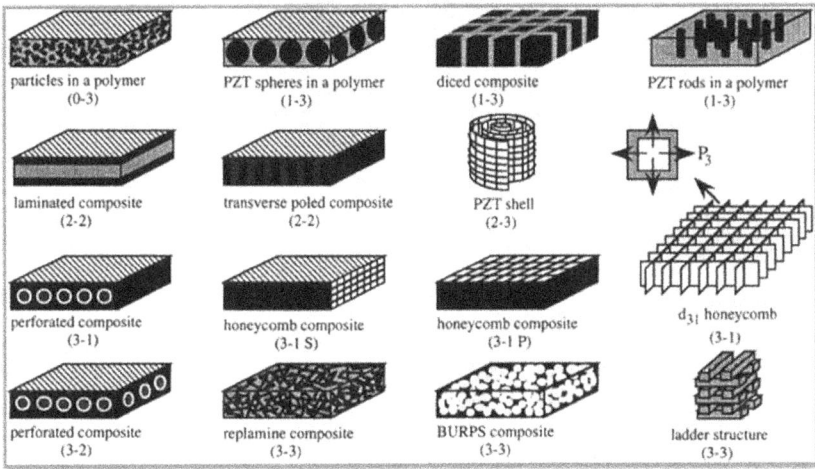

FIGURE 3.10 Schematic diagram showing classification of piezoelectric composites based on the their dimensional connectivity.

three-dimensionally continuous polymer/ceramic matrix. The properties of these ceramics greatly depend upon the ceramic filler particles, polymer/ceramic matrix as well as the synthesis technique. Composites possessing such a configuration are suitable for the formation of different shapes and deposition on surfaces. However, the composites must be poled by introducing an electric field above their coercive field in order to proliferate the piezoelectric capabilities. However, the greatest risk in doing so is the polymer matrix's poor permittivity and conductivity, which limits poling to areas that are close to the electrode. Eßlinger et al. [48] discussed a novel technique to pole the composites (PZT-epoxy and PZT-Polyurethene) in the desired direction, which utilizes the advantages of electrical conductivity of some polymers both in uncured or partially cured condition. The observation of non-zero piezoelectric charge constant of 28.6 ± 3.1 pC/N for PZT-epoxy and 35.5 ± 0.7 pC/N for PZT-polyurethene claims the existence of piezoelectricity in the composites. Highly flexible piezoelectric composite with 0–3 configuration was fabricated by Babu et al. [49] using PZT as the filler particles in polydimethylsiloxane (PDMS) matrix. This solution casted composite exhibited good piezoelectric properties with permittivity ~40; piezoelectric charge constant ~25 pC/N and voltage constant ~75 mV m/N (Figure 3.11). The superior flexibility of the composites in combination with optimum piezo properties suggested soft touch applications in a variety of transducers and sensors. Recently, 0–3 PZT-PDMS composites were synthesized by a low-cost and simple process involving stirring, spin coating and curing. Corona poling was used to align the dipoles with the direction of the applied electric field and enhance the material's ability to respond to pressure. Due to the alignment of the domains along the poling axis, the piezoelectric response improved after poling (electric field). The results show that a composite with 28 volume percent of polarized PZT particles can provide piezoelectric charge coefficients of 78.33

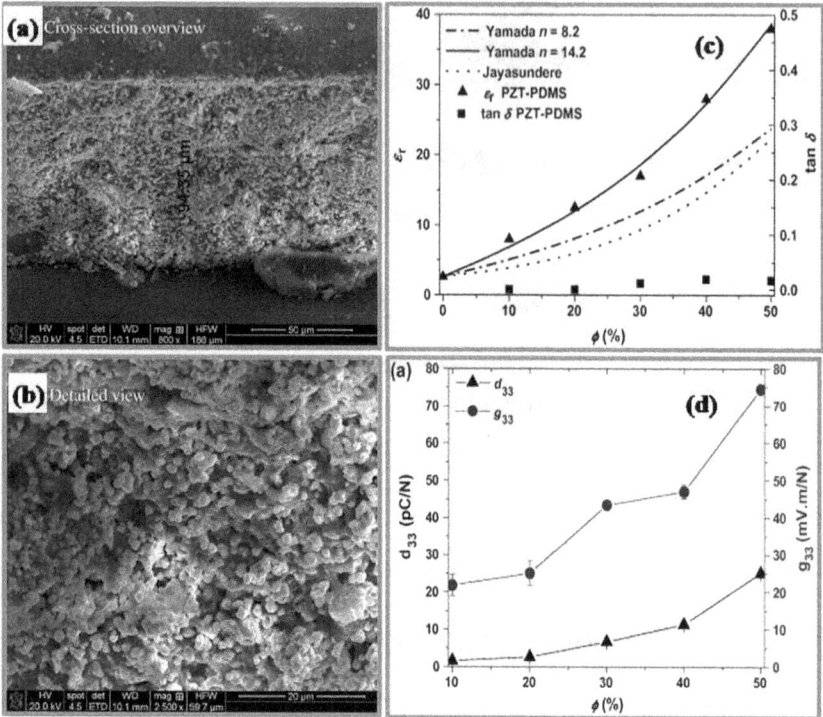

FIGURE 3.11 (a) Cross sectional view (SEM) and (b) detailed view of PZT-PDMS composite with 50% PZT of thickness ~100 μm; (c) dielectric permittivity of PZT-PDMS composites in comparison to Yamada and Jayasundere model; (d) variation of d_{33} and g_{33} as a function of PZT% in PZT-PDMS composites.

Source: [49].

pC/N, 10 MPa Young's modulus, and 10 relative permittivity. When finger pressure was applied, the thin film made from this mixture produced charges throughout the film that were proportional to the amount of pressure [50].

3.3.2   1–3 Piezo Composites

Ceramic fillers being oriented in one direction within the matrix material constitute 1–3 composite structure, which offers a wide range of applications in medical imaging transducers, nondestructive testing devices, underwater sonars and so forth. Depending on the spatial arrangement of the fibers and the relative orientation of the direction of poling of one phase with regard to another, 1–3 piezo composites can be divided into 15 varieties. There are three different types of characteristic choices based on the configurational arrangement of the fibers within the matrix material: square diagonal, square edge, and hexagonal. Similarly, three well-defined choices of relative poling directions between filler and matrix (unpoled, poling direction parallel or perpendicular to the fiber axis respectively) differentiates five distinct piezo composite types (Figure 3.12):

- Both the fiber and matrix phases were poled in the (+3) direction, parallel to the fiber axis.
- Anti-parallel (+3 and −3 orientations) alignment/poling of the matrix and fiber
- Fiber that is poled in a direction (+1 direction) that is perpendicular to the matrix (+3 direction)
- Matrix poled in a direction (+1 direction) perpendicular to the fiber phase (+3 direction)
- Matrix unpoled and fiber poled along its axis (+3 direction)

In this context, Gupta and Venkatesh in 2005 [51] made a theoretical study on the effect of relative orientation of the poling direction (between the filler and matrix phase) on the electromechanical response of 1–3 piezocomposites. Some of the results based on their study indicate that: (i) rational selection of poling alignment of the constituent phases improve unidirectional and bidirectional sensitivity; (ii) variation of dielectric, piezoelectric, elastic material characteristics with the

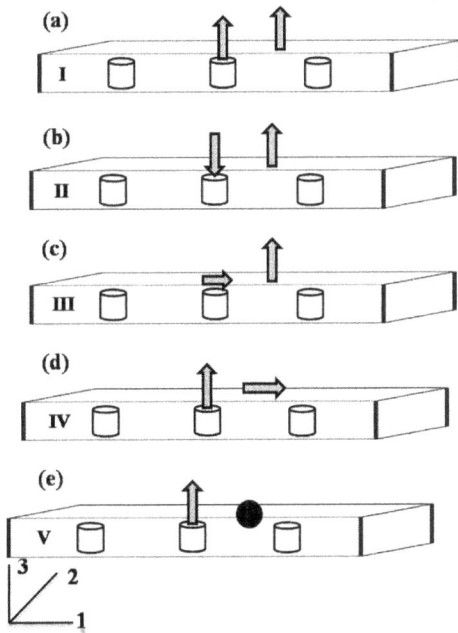

FIGURE 3.12 Schematic figure showing five distinct piezo composite types based on relative poling directions of filler and matrix.

fiber volume fraction is non linear while converging to the monolithic material limits at very high and low volume fractions; (iii) matrix-fiber interface are identified to be the primary regions where mechanical stresses accumulate and, thus, location and onset of failure in piezocomposites can be predicted; (iv) the highest electromechanical energy conversion can be obtained in piezocomposites in which both fiber and matrix phases are poled in a direction parallel to the fiber axis. Further, the work is extended by different researchers by making individual study on the effect of internal stresses and poling conditions on the properties of piezocomposites. Wang et al. [52] investigated on the influence of interfacial stresses on the functional property of PMN-PT single crystal filler in epoxy matrix. They studied the changes in domain configuration and stress distribution on poling the crystal rods using polarizing light microscopy and PFM. It was noticed that the stress generated due to the polymer filler interaction led to incomplete poling in the near-rod areas. High temperature poling significantly improved the piezoelectric properties of the composite. Lately, performance of 1–3 composites was studied under different poling conditions. The dielectric and electromechanical characteristics of these materials were significantly influenced by poling frequency (f), electric field amplitude (E_p), and cycle number (C) (Figure 3.13). Samples poled under optimum alternating current poling (f = 1.25 Hz, E_p = 2 kV/mm and C = 16) demonstrated an increase of d_{33}, dielectric constant and k_t of 21.4 percent, 13 percent and 10.7 percent respectively as compared to direct current poling [53].

3.3.3   3–3 PIEZO COMPOSITES

3–3 piezocomposites are composed of matrix of two different phases: one is a piezoelectric active phase, while the other is a polymer passive phase. Both phases interpenetrate completely so that each phase forms a three-dimensional network around the other. Connectivity of both the phases is highest in case of 3–3 composites as compared to the other two configurations. Due to effective stress transfer, such three-dimensional connection produces improved piezoelectric activity at low dielectric permittivity. With the same volume fraction of the piezoactive component, the 3–3 connectivity will perform electromechanically better than the 0–3 and 1–3 connectivities. Low density and stiffness, comparatively strong piezoelectric sensitivity, enhanced acoustic matching with water and human tissue, high mechanical strength for machining, and damping compliance – these are the main benefits of 3–3 connection. In spite of such advantages, experimental studies on 3–3 composites are rare, though mostly theoretical simulation works are reported. Bowen and Topolov in 2003 [54] modeled and compared the performances of 0–3 and 3–3 connected piezocomposites and provided important inputs for the prediction of piezoelectric sensitivity and property optimization. The model anticipated significant improvement in the piezosensitive parameters in $PbTiO_3$ based composite. The arrangement of ceramic component in all the samples was reported to affect electromechanical performance of the 3–3 configuration; 3–3–3

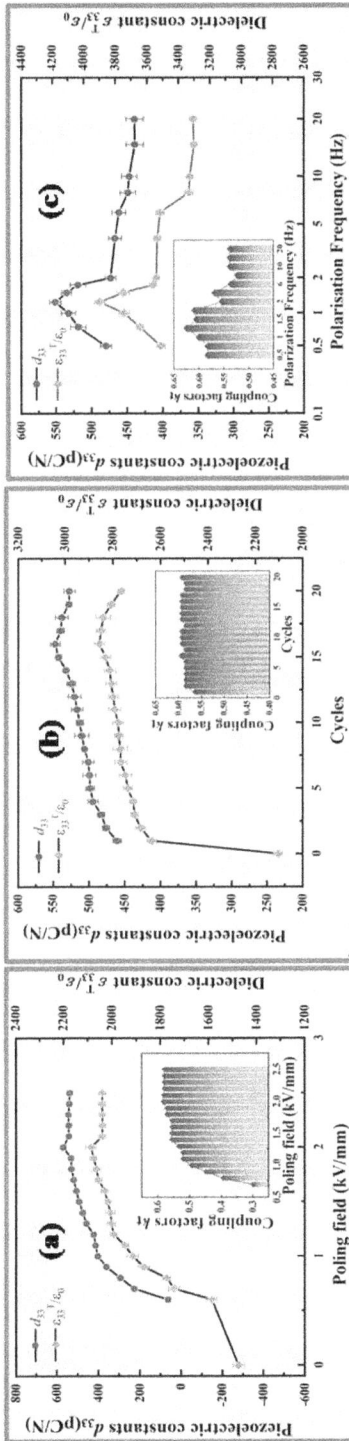

FIGURE 3.13 d_{33}, $(\varepsilon^{T}_{33})/\varepsilon_{0}$ and k_{t} of 1-3 piezocomposite at various (a) AC electric field amplitudes; (b) cycle number; (c) poling frequencies.

Source: [53].

piezoelectric polymer composites fabricated by an easy technique of immersing an ordered polymer mesh in to PZT suspension. Thereafter the mesh was allowed to undergo drying, pyrolysis, and sintering. The resulting PZT structures are porous when subjected to polymer injection to form composites. To investigate the effects of polarization direction and the amount of active piezoelectric phase on the dielectric and piezoelectric properties, the structures were sliced and polarized in various directions. Piezoelectric and voltage coefficients were discovered to be at their highest in a direction 45 degrees from the ceramic ligaments, despite the fact that the dielectric constant is higher in a direction parallel to the ceramic ligaments [55]. Recently, direct ink technique was adopted for producing 3-3 composites, since it is a straight-forward method for achieving accurate control of the structures. In this method, PZT scaffolds with varying porosity were initially prepared by direct ink writing and then 3-3 structures were realized by compounding PZT scaffolds with epoxy resin. It was noticed that Y-poled samples exhibited higher piezo performance than Z-poled structures (Figure 3.14). With a PZT content of 34.4 percent, the Y-poled samples showed voltage constant of 81.65 mV m/N and sensitivity as high as 9.81 V/N [56].

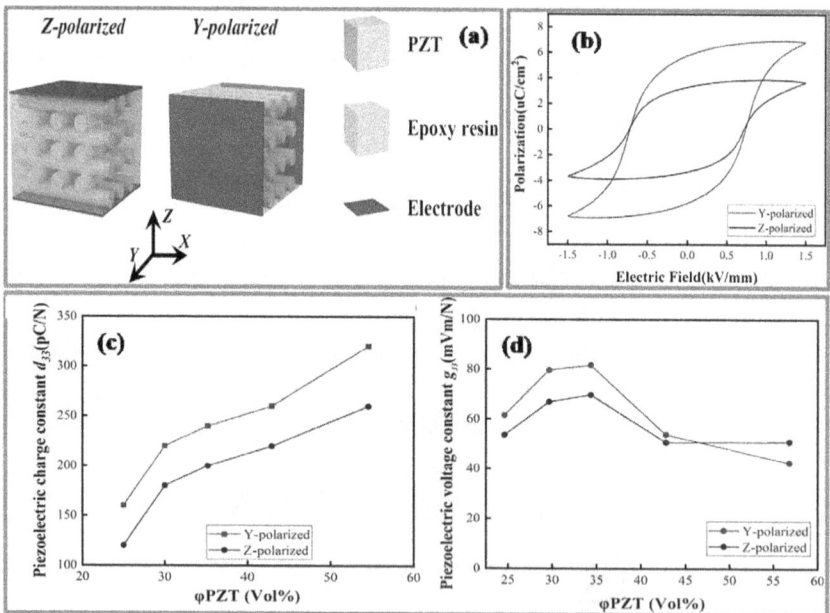

FIGURE 3.14 (a) Schematic diagram showing the polarization directions of 3-3 piezoelectric composites; (b) P-E loop of composites (PZT scaffolds in epoxy resin; (c) and (d) d_{33} and g_{33} as a function of PZT vol% respectively.

Source: [56].

3.4 BIO-INSPIRED

Simple, yet effective, adaptations are provided by biological systems to actualize particular functionality and deal with challenging situations. Finding solutions to many human problems can be accomplished via careful reflection and comprehension of these adaptations. Such an approach is called "bio-inspiration" and has been used as a common tool in recent years in applications such as material science, robotics, medicine, chemistry, and so on. This is a rapidly growing field and in this section we discuss a few examples of such bioinspired materials. A piezoelectric polymeric material was created in 2018 by Roca et al. [57] and used to 3D-print a new frequency selective sensor that was inspired by the tympanic membrane of a locust (Figure 3.15). This study is inspired by the insect's acoustic systems, which involve different mechanisms for detecting and processing sound. Some of them have tympanal ears, which are mostly made up of a membrane covering a chamber filled with fluid. The neural nerves (scolopidia) are suspended from this cavity. On an impression of any incoming sound, the membrane undergoes mechanical motion, which is then converted to electrical signals by the scolopedia and transferred to the central nervous system. The locust is one such insect having a pear-shaped tympanic membrane made up of a cuticle of graded thickness. This simple structure of graded membrane allows selective acoustic frequency response helping the insect in predation and swarming. Therefore, the authors utilized this smart and simple structure of locust's tympanum to design piezocomposite based on $BaTiO_3$ nanoparticles and polymerizable methacrylate group linked to bisphenol by ethylene oxide chains of different lengths. In another interesting work, piezoelectric composite generators were designed based on the structure of the sea sponge. The sea sponge is one of the uncomplicated multicellular organisms having an interesting composite structure made up of soft fibrils (sponging) embedded within hard skeletons (spicules) forming a three-dimensional porous configuration. Such an arrangement demonstrates a high toughness as well as more elasticity and was taken as an inspiration to model composites made of (Ba,Ca) $(Zr,Ti)O_3$ piezoceramic within an elastomer matrix. Theoretical simulations on this composite predicted remarkable improvement in piezo potential (open circuit voltage ~25 V; short circuit current ~550 nA cm^{-2} and instantaneous power density ~2.6 µW cm^{-2}) that is almost 16 times greater than conventional piezoelectric polymer composite. In addition, the composite generator demonstrated that strain-voltage efficiency under stretching was thirty times higher than the claimed state-of-the-art performance of several other energy harvesting devices [58]. Besides, an efficient bio-piezoelectric nanogenerator based on the swim bladder of Catla Catla fish (fresh sweet water fish) was fabricated by Ghosh et al. [59] in 2016. The fish swim bladder is one of the main waste products in fish processing comprising of well aligned collagen nano-fibrils. The nanogenerator fabricated out of these bladders generate an open circuited voltage of 10 V along with a short circuited current of 51 nA in response to a periodic compressive normal stress of 1.4 MPa (human finger). Apart from that, an output power density of 4.15 µW cm^{-2} and high electromechanical energy conversion efficiency of 0.3 percent, which was

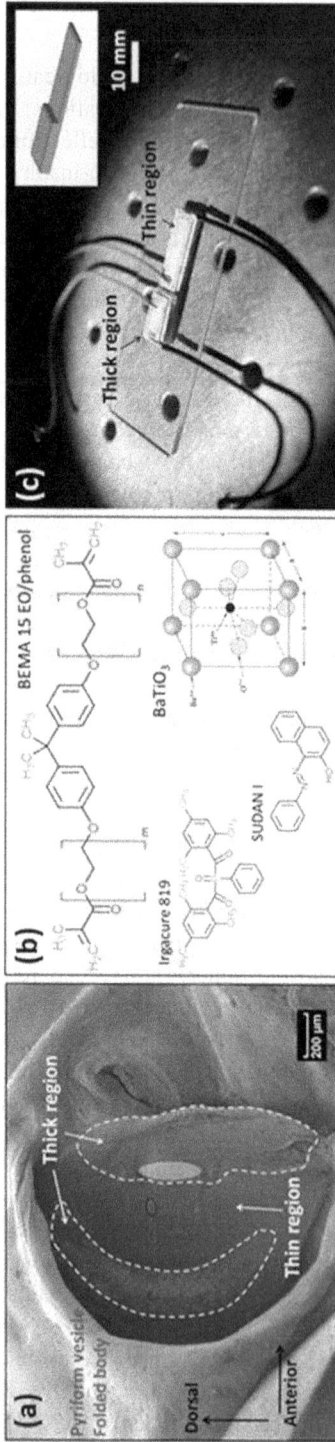

FIGURE 3.15 (a) SEM image of locust TM; (b) monomer, photoinitiator, dye and nanoparticle crystallographic structure; (c) locust TM bio-inspired 3D printed sensor and corresponding CAD file; (d) locust TM bio-inspired 3D printed sensor and corresponding CAD file.

Source: [57].

adequate enough to light up 50 blue, 22 green as well as red and white LEDs. Moreover, a multitude of biomaterials including onion skin, eggshell membrane, bacteriophages consisting of piezoelectric components such as vitamins, chitin, collagen fibrils and so forth. These materials are not only biodegradable and biocompatible but also flexible and durable. Thus, use of these bio-inspired natural materials to design piezoelectric nanogenerators will not on reduce the amount of biological wastes but also limit the amount of toxic e-wastes generated from electronic devices.

3.5 SUMMARY

This chapter provides a detailed overview of different types prospective piezoelectric materials and their recent developments. As discussed in preceding sections, ceramic materials (nano and bulk) with tailored morphologies and compositions exhibit superior piezoelectric performance as compared to other categories; however, brittleness is a major concern. To explore the full potentiality of the ceramic materials (nano formulation in the form of particles, rods, plates, thin films), they are embedded into a flexible polymer matrix forming composites. This strategy has attracted the attention of researchers and industry since it brings together the flexibility of the polymers and favorable piezoelectricity of ceramics. Further, the importance of bio-inspired materials is growing rapidly and, in the coming years, it will not only reduce the biological wastes but also limit toxic e-wastes from electronic industries. Overall, the discussions made in this text may guide young brains to develop novel materials with enhanced piezoelectric properties.

REFERENCES

1. Hao, J., Li, W., Zhai, J. and Chen, H., 2019. Progress in high-strain perovskite piezoelectric ceramics. *Materials Science and Engineering: R: Reports, 135*, pp. 1–57.
2. Jaffe, H., 1958. Piezoelectric ceramics. *Journal of the American Ceramic Society, 41*(11), pp. 494–498.
3. Donnelly, N.J., Shrout, T.R. and Randall, C.A., 2007. Addition of a Sr, K, Nb (SKN) combination to PZT (53/47) for high strain applications. *Journal of the American Ceramic Society, 90*(2), pp. 490–495.
4. Luo, J., Qiu, J., Zhu, K., Du, J., Pang, X. and Ji, H., 2011. Effects of the calcining temperature on the piezoelectric and dielectric properties of 0.55PNN-0.45PZT ceramics. *Ferroelectrics, 425*(1), pp. 90–97.
5. Du, J., Qiu, J., Zhu, K. and Ji, H., 2014. Enhanced piezoelectric properties of 0.55Pb $(Ni_{1/3}Nb_{2/3})O_3$–0.135PbZrO$_3$–0.315PbTiO$_3$ ternary ceramics by optimizing sintering temperature. *Journal of Electroceramics, 32*(2), pp. 234–239.
6. Birol, H., Damjanovic, D. and Setter, N., 2006. Preparation and characterization of $(K_{0.5}Na_{0.5})NbO_3$ ceramics. *Journal of the European Ceramic Society, 26*(6), pp. 861–866.

7. Singh, K., Lingwal, V., Bhatt, S.C., Panwar, N.S. and Semwal, B.S., 2001. Dielectric properties of potassium sodium niobate mixed system. *Materials Research Bulletin*, *36*(13–14), pp. 2365–2374.

8. Du, H., Li, Z., Tang, F., Qu, S., Pei, Z. and Zhou, W., 2006. Preparation and piezoelectric properties of $(K_{0.5}Na_{0.5})NbO_3$ lead-free piezoelectric ceramics with pressure-less sintering. *Materials Science and Engineering: B*, *131*(1–3), pp. 83–87.

9. Kosec, M. and Kolar, D., 1975. On activated sintering and electrical properties of NaKNbO3. *Materials Research Bulletin*, *10*(5), pp. 335–339.

10. Saito, Y., Takao, H., Tani, T., Nonoyama, T., Takatori, K., Homma, T., Nagaya, T. and Nakamura, M., 2004. Lead-free piezoceramics. *Nature*, *432*(7013), pp. 84–87.

11. Wang, T., Liao, Y., Wang, D., Zheng, Q., Liao, J., Xie, F., Jie, W. and Lin, D., 2019. Cycling-and heating-induced evolution of piezoelectric and ferroelectric properties of CuOdoped $K_{0.5}Na_{0.5}NbO_3$ ceramic. *Journal of the American Ceramic Society*, *102*(1), pp. 351–361.

12. Kanuru, S.R., Baskar, K. and Dhanasekaran, R., 2016. Synthesis, structural, morphological and electrical properties of NBT–BT ceramics for piezoelectric applications. *Ceramics International*, *42*(5), pp. 6054–6064.

13. Li, W.L., Cao, W.P., Xu, D., Wang, W. and Fei, W.D., 2014. Phase structure and piezoelectric properties of NBT–KBT–BT ceramics prepared by sol–gel flame synthetic approach. *Journal of Alloys and Compounds*, *613*, pp. 181–186.

14. Praharaj, S., Rout, D., Kang, S.J. and Kim, I.W., 2016. Large electric field induced strain in a new lead-free ternary $Na_{0.5}Bi_{0.5}TiO_3$-$SrTiO_3$-$BaTiO_3$ solid solution. *Materials Letters*, *184*, pp. 197–199.

15. Duraisamy, D. and Venkatesan, G.N., 2018. Compositionally driven giant strain and electrostrictive co-efficient in lead free NBT-BT-BFO system. *Applied Physics Letters*, *112*(5), p. 052903.

16. Park, S.E. and Shrout, T.R., 1997. Ultrahigh strain and piezoelectric behavior in relaxor based ferroelectric single crystals. *Journal of Applied Physics*, *82*(4), pp. 1804–1811.

17. Zhang, S., Li, F., Yu, F., Jiang, X., Lee, H.Y., Luo, J. and Shrout, T.R., 2018. Recent developments in piezoelectric crystals. *Journal of the Korean Ceramic Society*, *55*(5), pp. 419–439.

18. Trolier-McKinstry, S., Zhang, S., Bell, A.J. and Tan, X., 2018. High-performance piezoelectric crystals, ceramics, and films. *Annual Review of Materials Research*, *48*, pp. 191–217.

19. Tian, F., Liu, Y., Ma, R., Li, F., Xu, Z. and Yang, Y., 2021. Properties of PMN-PT single crystal piezoelectric material and its application in underwater acoustic transducer. *Applied Acoustics*, *175*, p. 107827.

20. Tazaki, R., Fu, D., Itoh, M., Daimon, M. and Koshihara, S.Y., 2009. Lattice distortion under an electric field in $BaTiO_3$ piezoelectric single crystal. *Journal of Physics: Condensed Matter*, *21*(21), p. 215903.

21. Zheng, L., Huo, X., Wang, R., Wang, J., Jiang, W. and Cao, W., 2013. Large size lead-free $(Na,K)(Nb,Ta)O_3$ piezoelectric single crystal: growth and full tensor properties. *CrystEngComm*, *15*(38), pp. 7718–7722.

22. Sun, R., Zhao, X., Zhang, Q., Fang, B., Zhang, H., Li, X., Lin, D., Wang, S. and Luo, H., 2011. Growth and orientation dependence of electrical properties of $0.92Na_{0.5}Bi_{0.5}TiO_3$–$0.08K_{0.5}Bi_{0.5}TiO_3$ lead-free piezoelectric single crystal. *Journal of Applied Physics*, *109*(12), p. 124113.

23. Ge, W., Liu, H., Zhao, X., Li, X., Pan, X., Lin, D., Xu, H., Jiang, X. and Luo, H., 2009. Orientation dependence of electrical properties of $0.96Na_{0.5}Bi_{0.5}TiO_3$–$0.04BaTiO_3$ lead-free piezoelectric single crystal. *Applied Physics A*, *95*(3), pp. 761–767.
24. Moon, K.S., Rout, D., Lee, H.Y. and Kang, S.J.L., 2011. Solid state growth of $Na_{1/2}Bi_{1/2}TiO_3$–$BaTiO_3$ single crystals and their enhanced piezoelectric properties. *Journal of Crystal Growth*, *317*(1), pp. 28–31.
25. Chen, C., Zhao, X., Wang, Y., Zhang, H., Deng, H., Li, X., Jiang, X., Jiang, X. and Luo, H., 2016. Giant strain and electric-field-induced phase transition in lead-free $(Na_{0.5}Bi_{0.5})TiO_3$-$BaTiO_3$-$(K_{0.5}Na_{0.5})NbO_3$ single crystal. *Applied Physics Letters*, *108*(2), p. 022903.
26. Yang, S., Li, J., Liu, Y., Wang, M., Qiao, L., Gao, X., Chang, Y., Du, H., Xu, Z., Zhang, S. and Li, F., 2021. Textured ferroelectric ceramics with high electromechanical coupling factors over a broad temperature range. *Nature Communications*, *12*(1), pp. 1–10.
27. Li, P., Zhai, J., Shen, B., Zhang, S., Li, X., Zhu, F. and Zhang, X., 2018. Ultrahigh piezoelectric properties in textured (K, Na)NbO₃-based lead-free ceramics. *Advanced Materials*, *30*(8), p. 1705171.
28. Maurya, D., Pramanick, A., An, K. and Priya, S., 2012. Enhanced piezoelectricity and nature of electric-field induced structural phase transformation in textured lead-free piezoelectric $Na_{0.5}Bi_{0.5}TiO_3$-$BaTiO_3$ ceramics. *Applied Physics Letters*, *100*(17), p. 172906.
29. Maurya, D., Zhou, Y., Wang, Y., Yan, Y., Li, J., Viehland, D. and Priya, S., 2015. Giant strain with ultra-low hysteresis and high temperature stability in grain oriented lead-free $K_{0.5}Bi_{0.5}TiO_3$-$BaTiO_3$-$Na_{0.5}Bi_{0.5}TiO_3$ piezoelectric materials. *Scientific Reports*, *5*(1), pp. 1–8.
30. Wang, Z.L. and Song, J., 2006. Piezoelectric nanogenerators based on zinc oxide nanowire arrays. *Science*, *312*(5771), pp. 242–246.
31. Kim, K.H., Kumar, B., Lee, K.Y., Park, H.K., Lee, J.H., Lee, H.H., Jun, H., Lee, D. and Kim, S.W., 2013. Piezoelectric two-dimensional nanosheets/anionic layer heterojunction for efficient direct current power generation. *Scientific Reports*, *3*(1), pp. 1–6.
32. Fortunato, M., Chandraiahgari, C.R., De Bellis, G., Ballirano, P., Soltani, P., Kaciulis, S., Caneve, L., Sarto, F. and Sarto, M.S., 2018. Piezoelectric thin films of ZnO-nanorods/nanowalls grown by chemical bath deposition. *IEEE Transactions on Nanotechnology*, *17*(2), pp. 311–319.
33. Chen, X., Xu, S., Yao, N. and Shi, Y., 2010. 1.6 V nanogenerator for mechanical energy harvesting using PZT nanofibers. *Nano Letters*, *10*(6), pp. 2133–2137.
34. Karaki, T., Yan, K., Miyamoto, T. and Adachi, M., 2007. Lead-free piezoelectric ceramics with large dielectric and piezoelectric constants manufactured from BaTiO3 nano-powder. *Japanese Journal of Applied Physics*, *46*(2L), p. L97.
35. Park, K.I., Son, J.H., Hwang, G.T., Jeong, C.K., Ryu, J., Koo, M., Choi, I., Lee, S.H., Byun, M., Wang, Z.L. and Lee, K.J., 2014. Highly-efficient, flexible piezoelectric PZT thin film nanogenerator on plastic substrates. *Advanced Materials*, *26*(16), pp. 2514–2520.
36. Cernea, M., Galca, A.C., Cioangher, M.C., Dragoi, C. and Ioncea, G., 2011. Piezoelectric BNT-$BT_{0.11}$ thin films processed by sol–gel technique. *Journal of Materials Science*, *46*(17), pp. 5621–5627.

37. Kim, B.Y., Seo, I.T., Lee, Y.S., Kim, J.S., Nahm, S., Kang, C.Y., Yoon, S.J., Paik, J.H. and Jeong, Y.H., 2015. High-performance $(Na_{0.5}K_{0.5})NbO_3$ thin film piezoelectric energy Harvester. *Journal of the American Ceramic Society, 98*(1), pp. 119–124.

38. Qin, W., Li, T., Li, Y., Qiu, J., Ma, X., Chen, X., Hu, X. and Zhang, W., 2016. A high power ZnO thin film piezoelectric generator. *Applied Surface Science, 364*, pp. 670–675.

39. Ounaies, Z., Young, J.A. and Harrison, J.S., 1999. An overview of the piezoelectric phenomenon in amorphous polymers. In *Field Responsive Polymers, ACS Symposium Series*, pp. 88–103.

40. Kaura, T., Nath, R. and Perlman, M.M., 1991. Simultaneous stretching and corona poling of PVDF films. *Journal of Physics D: Applied Physics, 24*(10), p. 1848.

41. Gomes, J., Nunes, J.S., Sencadas, V. and Lanceros-Méndez, S., 2010. Influence of the β-phase content and degree of crystallinity on the piezo and ferroelectric properties of poly (vinylidene fluoride). *Smart Materials and Structures, 19*(6), p. 065010.

42. Ibtehaj, K., Jumali, M.H.H., Al-Bati, S., Ooi, P.C., Al-Asbahi, B.A. and Ahmed, A.A.A., 2022. Effect of β-chain alignment degree on the performance of piezoelectric nanogenerator based on poly(vinylidene fluoride) nanofiber. *Macromolecular Research, 30*(3), pp. 172–182.

43. Cauda, V., Stassi, S., Bejtka, K. and Canavese, G., 2013. Nanoconfinement: an effective way to enhance PVDF piezoelectric properties. *ACS Applied Materials & Interfaces, 5*(13), pp. 6430–6437.

44. Farahani, A., Zarei-Hanzaki, A., Abedi, H.R., Haririan, I., Akrami, M., Aalipour, Z. and Tayebi, L., 2021. An investigation into the polylactic acid texturization through thermomechanical processing and the improved d_{33} piezoelectric outcome of the fabricated scaffolds. *Journal of Materials Research and Technology, 15*, pp. 6356–6366.

45. Yanase, T., Hasegawa, T., Nagahama, T. and Shimada, T., 2012. Fabrication of piezoelectric polyurea films by alternating deposition. *Japanese Journal of Applied Physics, 51*(4R), p. 041603.

46. Newnham, R.E., Skinner, D.P. and Cross, L.E., 1978. Connech//ard AeZOelecic-O/ C-Georcomposites. *Materials Research Bulletin, 13*, pp. 525–536.

47. Pilgrim, S.M., Newnham, R.E. and Rohlfing, L.L., 1987. An extension of the composite nomenclature scheme. *Materials Research Bulletin, 22*(5), pp. 677–684.

48. Eßlinger, S., Geller, S., Hohlfeld, K., Gebhardt, S., Michaelis, A., Gude, M., Schönecker, A. and Neumeister, P., 2018. Novel poling method for piezoelectric 0–3 composites and transfer to series production. *Sensors and Actuators A: Physical, 270*, pp. 231–239.

49. Babu, I. and de With, G., 2014. Highly flexible piezoelectric 0–3 PZT–PDMS composites with high filler content. *Composites Science and Technology, 91*, pp. 91–97.

50. Sappati, K.K. and Bhadra, S., 2020. Flexible piezoelectric 0–3 PZT-PDMS thin film for tactile sensing. *IEEE Sensors Journal, 20*(9), pp. 4610–4617.

51. Kar-Gupta, R. and Venkatesh, T.A., 2005. Electromechanical response of 1–3 piezoelectric composites: Effect of poling characteristics. *Journal of Applied Physics, 98*(5), p. 054102.

52. Wang, C., Sun, E., Liu, Y., Zhang, R., Yang, B. and Cao, W., 2016. Structural deformation of $0.74Pb(Mg_{1/3}Nb_{2/3})O_3$-$0.26PbTiO_3$ single crystal in 1–3 composites

due to interface stresses and poling procedure optimization. *Journal of Applied Physics*, *120*(12), p. 124104.

53. Ma, J., Zhu, K., Huo, D., Shen, B., Liu, Y., Qi, X., Sun, E. and Zhang, R., 2021. Performance enhancement of 1–3 piezoelectric composite materials by alternating current polarising. *Ceramics International*, *47*(13), pp. 18405–18410.

54. Bowen, C.R. and Topolov, V.Y., 2003. Piezoelectric sensitivity of $PbTiO_3$-based ceramic/polymer composites with 0–3 and 3–3 connectivity. *Acta Materialia*, *51*(17), pp. 4965–4976.

55. Sharifi Olyaei, N., Mohebi, M.M. and Kaveh, R., 2017. Directional properties of ordered 3-3 piezocomposites fabricated by sacrificial template. *Journal of the American Ceramic Society*, *100*(4), pp. 1432–1439.

56. Li, J., Yan, M., Zhang, Y., Li, Z., Xiao, Z., Luo, H., Yuan, X. and Zhang, D., 2022. Optimization of polarization direction on 3D printed 3-3 piezoelectric composites for sensing application. *Additive Manufacturing*, *58*, p. 103060.

57. Domingo-Roca, R., Tiller, B., Jackson, J.C. and Windmill, J.F.C., 2018. Bio-inspired 3D-printed piezoelectric device for acoustic frequency selection. *Sensors and Actuators A: Physical*, *271*, pp. 1–8.

58. Zhang, Y., Jeong, C.K., Yang, T., Sun, H., Chen, L.Q., Zhang, S., Chen, W. and Wang, Q., 2018. Bioinspired elastic piezoelectric composites for high-performance mechanical energy harvesting. *Journal of Materials Chemistry A*, *6*(30), pp. 14546–14552.

59. Ghosh, S.K. and Mandal, D., 2016. Efficient natural piezoelectric nanogenerator: electricity generation from fish swim bladder. *Nano Energy*, *28*, pp. 356–365.

4 Synthesis/Fabrication Techniques of Piezoelectric Materials

4.1 SYNTHESIS TECHNIQUES

Synthesis of piezoelectric materials is a very important step in modulating the particle size, morphology, formation of secondary phases, density, and so forth, for specific applications. A schematic diagram listing almost all the important synthesis techniques is displayed in Figure 4.1.

4.1.1 POWDER PROCESSING

Powder processing is one of the most accepted techniques for material synthesis apart from melt casting or vapor deposition. The basic steps of the powder processing route include preparation of powder→shape forming→sintering at high temperature→finished product. The preparation of ceramic powder with optimum particle size is one of the crucial parameters for further processing and obtaining the final material. It requires a greater degree of process control to obtain the desired chemical purity, microstructure, homogeneous grain-grain boundary distribution, defect formation, dispersion of phases, fine particle size to facilitate sintering, and so forth. Hence, choice of an appropriate processing method is highly essential. In this section, a few of the powder synthesis routes are discussed to acknowledge the wide range of prospective piezoelectric materials. The pros and cons of different powder processing techniques are given in Table 4.1.

4.1.1.1 Solid-State Reaction

Solid-state reaction is a familiar method of synthesizing polycrystalline materials from solid reagents. Usually, the solids do not react at ambient temperature over normal time scales. Hence, for the reaction to occur at an appreciable rate, a relatively high temperature (1000–1500 °C) is often employed. The important factors affecting such reactions include reactivity, free energy, morphology, and the surface area of the solid reagents. Apart from those, other conditions such as pressure, temperature, and environment also control the rate of solid-state reaction.

DOI: 10.1201/9781003317289-4

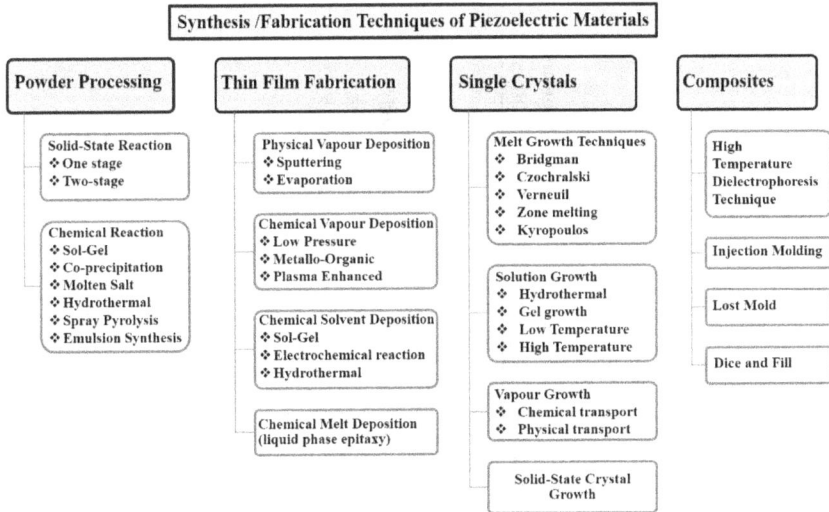

FIGURE 4.1   Schematic of piezoelectric material synthesis techniques.

The different steps involved in the conventional technique are mixing → solid state reaction → milling of powders → compacting the powders to obtain pellets. The final product obtained in this process may be either structured or random. More often aggregates or hard agglomerations are formed in this process, which requires further comminution to reduce the size of the particles at least to micrometer level. However, narrowing down the particle size is sometimes tedious and energy-intensive for hard substances. This results in a broad particle size distribution with lack of homogeneity and purity. Even the use of high calcination temperatures may also lead to the release of hazardous substances. For instance, volatile oxides such as PbO is released in $Pb(Zr,Ti)O_3$ synthesis. The presence of Pb in the ecosystem is very alarming and is listed in hazardous material category. In spite of these limitations, the solid-state reaction route has been explored both on research and industrial scale due to the simplicity of the process and low cost. Typically, the technique involves single stage or two stages, depending on the number of calcination steps (a schematic comparison between them is shown in Figure 4.2).

4.1.1.1.1   One Stage Solid-State Reaction

This is a simple approach in which all the reagents in powder form are mixed together and subjected to different stages of synthesis. Such a technique is most suited for the preparation of lead-free materials. Wang et al. [1] synthesized a multicomponent system consisting of $Bi_{0.5}Na_{0.5}TiO_3$ (BNT), $BaTiO_3$ (BT), and $Na_{0.73}Bi_{0.09}NbO_3$ (NBN) by a single stage solid-state reaction route. They initially mixed all the precursor materials in a stoichiometric ratio and then ball milled with zirconia balls in an alcohol medium for about 12 hours. The resulting slurry was

TABLE 4.1
Pros and cons of different powder processing techniques

Synthesis method	Solid-state Reaction	Sol-gel	Co-precipitation	Hydrothermal	Spray Pyrolysis	Emulsion Synthesis
State of development	commercial	R&D	commercial	demonstration	demonstration	demonstration
Compositional control	poor	excellent	good	excellent	excellent	excellent
Morphology control	poor	moderate	moderate	good	excellent	excellent
Powder reactivity	poor	good	good	good	good	good
Particle size (nm)	>1000	>10	>10	>100	>10	>100
Purity (%)	<99.5	>99.9	>99.5	>99.5	>99.9	>99.9
Agglomeration	moderate	moderate	high	low	low	low
Calcination step	yes	yes	yes	no	no	yes
Milling step	yes	yes	yes	no	yes	yes
Costs	low-moderate	moderate-high	moderate	moderate	high	moderate

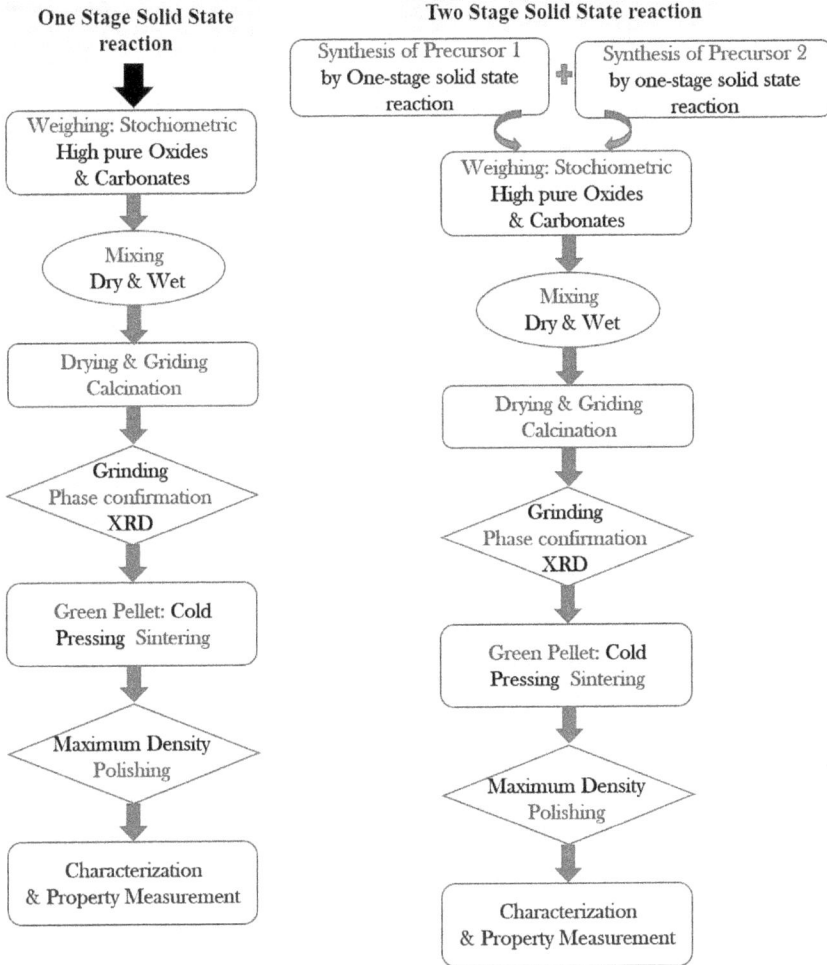

FIGURE 4.2   Schematic comparison of one stage and two stage solid state reactions.

dried at 80 °C, sieved and pressed into blocks before calcination. BNT-BT mixture was pre-calcined at 860 °C for 4 hours while NBN was calcined for 2 hours. The resulting powders were again milled for 12 hours and dried immediately. Further, the dried powders were isostatically pressed at 150 MPa and sintered at 1150–1170 °C for 3 hours. The samples exhibited uniform microstructure with appreciable dielectric and ferroelectric properties, though some secondary phases were detected. Some earlier studies focused on the kinetics of the reaction process [2]. Brzozowski and Castro [3] in 2000 proposed modifications in the kinetics of powder processing to obtain titanates free of secondary phases at a lower temperature. They suggested a mechanochemical activation in the synthesis of $BaTiO_3$ to limit the formation of secondary phases. Usually, BT is fabricated by the reaction between $BaCO_3$ and TiO_2. In the initial stage, the formation of $BaTiO_3$ depends on the decomposition of $BaCO_3$ which is promoted by the catalytic action of TiO_2. In the final stage, the creation of BT is governed by the diffusion of Ba through the barium titanate layer generating secondary phases. However, rigorous milling could increase the reactivity of barium carbonate refraining its decomposition as well as barium diffusion to control the course of reaction. Apart from the reaction kinetics between the intermediates, interactions between the reactant molecules and solvent are also important. Recently, a study was intended towards the role of calcinations, milling and mixing solvent (2-propanol) on the solid-state reaction between $BaCO_3$ and TiO_2. It was found that the size of the precursor particles reduced as a function of milling time from 18 μm for 2 hours of milling time to 1.2 μm for 24 hours of milling time. Again, the particle size increased as a function of thermal treatment temperature. Outcomes of the study suggest that intermediate steps of reaction between $BaCO_3$, TiO_2 as well as the dissociation of $BaCO_3$ were controlled by the dissociated and un-dissociated molecules of 2-propanol resulting in the scarcity of the functional groups on the $BaTiO_3$ particle surface at high temperatures [4]. This study not only provides the understanding of reaction mechanism at the structural level, but also evaluates the surface structure of $BaTiO_3$ particles. A single-stage solid-state reaction may also involve two calcinations or sintering steps to ensure better reaction among the precursors, to reduce the formation of secondary phases and obtain uniform particle size distribution. In this connection, Kainz et al. [5] followed a two-step calcination process in the synthesis of BNT-$Ba_{0.5}K_{0.5}TiO_3$ (BKT) piezoelectric ceramics and provided an insight into the formation process of the ceramics. The dried powders obtained after ball milling were sieved and then subjected to two-step calcination around 650 °C and 900 °C for 3 and 2 hours, respectively, followed by additional milling prior to sintering. The calcinations reaction of BNT-BKT was identified to be a multistep and endothermic process and, hence, studied by thermogravimetric analysis. The formation mechanism was proposed in terms of the following equations:

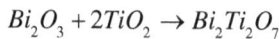

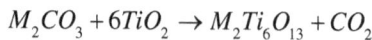

$$Bi_2O_3 + M_2CO_3 + TiO_2 \rightarrow 4Bi_{0.5}M_{0.5}TiO_2 + CO_2 \left(M = Na, K\right)$$

$$Bi_2O_3 + 2TiO_2 \rightarrow Bi_2Ti_2O_7$$

$$M_2CO_3 + 6TiO_2 \rightarrow M_2Ti_6O_{13} + CO_2$$

The first equation describes the formation of the perovskites while the second and third equations depict the formation of secondary phases. These two intermediates again interact with each other resulting in stoichiometric perovskites and titania as explained by the equation below:

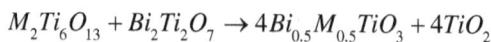

$$M_2Ti_6O_{13} + Bi_2Ti_2O_7 \rightarrow 4Bi_{0.5}M_{0.5}TiO_3 + 4TiO_2$$

Thus, it can be said that the required perovskite is formed by the reaction of the precursor materials and also the intermediate phases.

4.1.1.1.2   Two-Stage Solid-State Reaction

Typically, a two solid-stage reaction technique is followed while synthesizing lead-based piezoceramics $(Pb(Zr_{1-x}Ti_x)O_3$, $Pb(Mg_{1/3}Nb_{2/3})O_3$, $Pb(Yb_{1/2}Ta_{1/2})O_3$ and so on) which involve the formation of the pyrochlore phase. This is mostly formed during the early steps of one-stage reaction between the mixed oxides. Pyrochlore is parasitic, which not only degrades the electrical properties and lowers the dielectric constant of the Pb-ceramics, but also leads to less densification. Considering the detrimental effect of pyrochlore for application purposes, several researchers have made effort to eliminate the formation of this phase, but the most prominent one is the two-stage solid-state reaction. Swartz and Shrout [6] in 1982 introduced the columbite route (another name of the two-stage reaction) for the first time to synthesize $Pb(Mg_{1/3}Nb_{2/3})O_3$ (PMN) with good reproducibility. In this reaction path, MgO and Nb_2O_5 were precalcined to form $MgNb_2O_6$ (columbite) followed by reaction with PBO. The resulting PMN powder had very low pyrochlore content. To resolve the low reactivity problem of MgO, Liou et al. [7] modified the simple columbite route by directly pressing the mixture of $MgNb_2O_6$ and PbO into pellets and the sintering the pellets to produce PMN. Apparently, Bruno et al. [8] used a Ti-modified columbite route to synthesize (1-x)PMN-xPT solid solution. This procedure involves the preparation of columbite precursor via polymeric precursor route followed by solid-state reaction with PbO to obtain the final product. Samples prepared using this method exhibited high chemical and microstructural homogeneity. The two-stage technique was also adopted by Amer et al. [9] as an alternative to conventional single-stage solid-state reaction for synthesizing $Pb(Zr_{1-x}Ti_x)O_3$ (PZT) with better densification and enhanced electrical properties. In the first step, they prepared $(Zr_{1-x}Ti_x)O_3$ (ZTO) powder accompanied by reaction with PbO in the second step. The reactions are as follows:

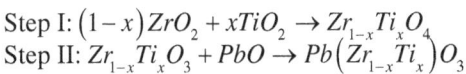

Step I: $(1-x)ZrO_2 + xTiO_2 \rightarrow Zr_{1-x}Ti_xO_4$
Step II: $Zr_{1-x}Ti_xO_3 + PbO \rightarrow Pb\left(Zr_{1-x}Ti_x\right)O_3$

Such a methodology refrains the occurrence of intermediate reactions and in turn restricts the formation of pyrochlore phase. In addition to that, it allows the formation of denser and homogeneous PZT ceramic at relatively low sintering temperature ~950 °C which is advantageous from the point of view of Pb volatilization. Again, PZT samples synthesized by two stage reaction exhibited better dielectric behavior than those prepared by the conventional method. Eventually, many other research groups also followed the improved two stage solid stage reaction route to synthesize other Pb based materials and could successfully eliminate or reduce the limitations associated with the one-stage reaction route [10-13].

4.1.1.2 Chemical Synthesis

Chemical synthesis refers to the synthesis methodology in which one or more chemical reactions take place to transform the reactants or pre-cursors into a final product. Apart from being a simple technique, chemical synthesis of mixed oxide powders promotes purity and chemical homogeneity. In addition, blending of the starting materials in the solution state ensures relatively low processing temperatures and production of fine particles. An overview of different chemical synthesis routes are displayed in Figure 4.3.

4.1.1.2.1   Sol-Gel

The sol-gel route is one of the effective synthesis approaches that ensure better control of the reactions, and homogeneity and high purity of the powders (metal oxides, carbides and nitrides) than other conventional processing methods. In this process, a reaction among the liquid precursors, usually colloidal solution (mainly by hydrolysis and condensation) leads to the formation of a sol which, again after a series of chemical reactions, turns into a network structure called "gel." The gel formed is then subjected to calcinations in order to obtain the final product. The sol-gel process can be briefly summarized in the following steps: (a) preparation of "sol" by hydrolysis and partial condensation of alkoxide; (b) creation of "gel" by polycondensation in the form of metal-oxo-metal or metal-hydroxy-metal bonds;

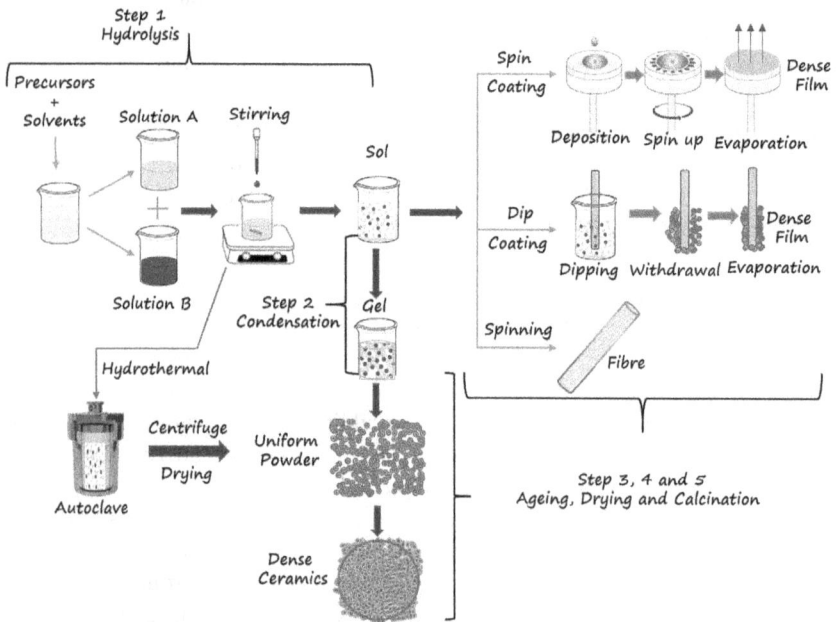

FIGURE 4.3   Overview of different chemical synthesis routes for powder and thin films.

(c) aging within the gel network often followed by shrinking and expulsion of the solvent; (d) formation of a dense "xerogel" on drying, which is mainly due to the collapse of the porous network or aerogel (may be termed supercritical drying); and (e) elimination of metal–OH groups via high temperature calcinations [14]. Apart from the chemistry of gel formation, other factors such as rate of evaporation during gelation, heat treatment procedure can be regulated to realize desirable chemical and structural features in the powder samples. This technique may not be suitable for the production of bulk materials but can be used conveniently for nano/ micro sized particles and thin films. Literature reports the synthesis of both Pb and non-Pb piezo materials. PZT thin film cantilever beams prepared by this remote technique were tested for the mechanical response using converse piezoelectric effect by Luginbuhl et al. [15] way back in 1996. Mechanical deflections measured by standard interferometry method yield a transverse piezoelectric coefficient d_{31} = –24.72 X 10^{-12} C/m. Gradually, the conventional technique was modified by many researchers. In the recent past, Kholin et al. [16], used a novel hybrid sol-gel approach to prepare composite PZT films of thickness 1–20 µm at lower sintering temperatures <600 °C. The technique includes a dip-coating method to deposit films (>5 µm) from sol-gel solution dispersed with PZT powders. The films of thickness >5 µm could be obtained via sedimentation of ceramic powders onto Pt-coated Si substrates followed be multiple infiltration of the powders with sol-gel solution. Optimum electrical properties with ε = 2500 and 1350; saturation polarization = 35 and 40 µC/cm^2 were obtained for dip coating and sedimentation respectively. In a work by Cernea et al. [17], lead-free barium titanate was doped in situ with 5.5 mol% of Ce via sol-gel process employing barium acetate, Ti(IV) isopropoxide and Ce(III) acetylacetonate as the precursor materials. The microstructure of the dried gel exhibited granular structure (~140 nm) with agglomerations. Persistence of OH–groups owing to the strong $Ti-OH$ bond up to temperatures as high as 720 °C were found to be responsible for the agglomeration. The as-prepared gel powder crystallized to perovskite barium titanate after air sintering at 1100 °C for 3 hours. Dielectric properties of the sample were almost frequency independent (100 Hz–1 MHz) and a high dielectric constant of 10130 and low loss of 0.018 was obtained at 100 Hz. Another group also employed the sol-gel method to synthesize 8 mol% $BaTiO_3$ doped $Na_{0.5}Bi_{0.5}TiO_3$ and studied the evolution of structure and microstructure of the precursor gel at different temperatures. The gel crystallized into perovskite phase with particle size ~30 nm when heated at ~600 °C. The as-prepared ceramic powder exhibited adequate electrical and piezoelectric properties with a maximum dielectric constant ~4000–4500; loss ~0.02–0.03 and a high mechanical quality factor ~500. These parameters are quite essential from the point of view of piezoelectric applications [18]. Besides, Dashtizad et al. [19] improved the piezoelectric properties of PVDF fibers by forming a composite with sol-gel synthesized $BaTiO_3$-Ag nanoparticles. These nanoparticles acted as the nucleation agents for promoting the formation of piezo-active β-phase. At an optimum concentration of 0.8 wt% $BaTiO_3$, a maximum amount of β-phase and an output voltage of 1.48(26) mV were achieved. Again, Ag particle could further enhance the amount of piezo-active phase of PVDF and improved the output voltage to 1.78(12) mV.

4.1.1.2.2 Co-Precipitation

Co-precipitation is one of the uncomplicated, industrially and economically viable synthesis routes, especially relevant for oxide materials. It involves the preparation of an aqueous solution of the starting agents and an appropriate precipitating agent. The main focus of this methodology is to synthesize multicomponent systems via the formation of intermediate precipitates typically hydrous oxides or oxalates. Besides, it also aims to form an exhaustive mixture of components during precipitation and maintain chemical homogeneity on calcination. Some of the advantages of co-precipitation include narrow particle size distribution in comparison to other wet chemical techniques, morphology control by the introduction of capping agents. However, the process is associated with some of the drawbacks like continuous washing, drying, and calcinations to obtain the pure phase. In spite of encouraging results, this method of synthesis is not adopted by many. Few years back, in 2007, Simon-Seveyrat et al. [20] compared the dielectric and piezoelectric properties of $BaTiO_3$ ceramics prepared by solid-state reaction and oxalate co-precipitation method. In the oxalate route, an aqueous solution of $Ti(OC_4H_9)_4$ is made with oxalic acid. A precipitate of $Ti(OH)_4$ is formed, which again reacts with oxalic acid, giving rise to soluble $TiOC_2O_4$. After a thermodynamically stable solution of Ti is formed, barium acetate is slowly added to obtain a $BaTiO(C_2O_4)_2.4H_2O$ (a double oxalate). The different steps involved in this co-precipitation process are summarized in the form of reactions as:

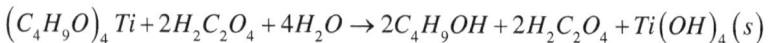

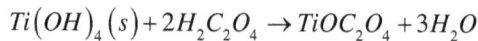

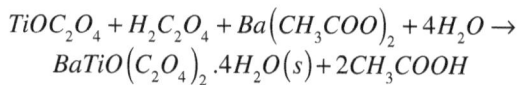

$$\left(C_4H_9O\right)_4 Ti + 2H_2C_2O_4 + 4H_2O \rightarrow 2C_4H_9OH + 2H_2C_2O_4 + Ti\left(OH\right)_4 (s)$$

$$Ti\left(OH\right)_4 (s) + 2H_2C_2O_4 \rightarrow TiOC_2O_4 + 3H_2O$$

$$TiOC_2O_4 + H_2C_2O_4 + Ba\left(CH_3COO\right)_2 + 4H_2O \rightarrow$$
$$BaTiO\left(C_2O_4\right)_2 .4H_2O(s) + 2CH_3COOH$$

The precursor powder is further calcined at 650 °C for 10 hours and again treated at 800 °C for 4 hours to regulate grain growth. These calcinations temperatures are much less than that required for solid state reaction (900 °C). In addition, no intermediate phases including Ba_2TiO_4 or $BaCO_3$ is created during heat treatment. This chemical process gave a good yield of fine grains with an adequate dielectric constant and a room temperature $d_{33} = 260$ pC/N, which is quite higher than solid-state reaction. On a similar note, pure and Nd doped ZnO nano rods synthesized by this method demonstrated remnant polarization and a coercive field of 0.064 µC/cm² and 6.04 kV/cm, respectively. Nanogenerators constructed out of those nanorods could produce an output voltage of 31 V as a result of high piezoelectric coefficient of 512 pm/V and reduced leakage current. The authors of this article recommend the use of Nd doped ZnO for pressure sensors and energy harvesting devices owing to its excellent piezo-performance [21]. In a recent study, a group of researchers synthesized hydroxypatite (Hap)/TiO$_2$ composites by a combination of ultrasound assisted sol-gel and co-precipitation techniques. The aim was to induce piezoelectricity in HAp owing to its use as a bioactive ceramic to replace hard tissue in the body. To their expectation, HAp obtained from the soft chemical process displayed ferroelectric domains, and the piezoelectric coefficient was measured to be 22.2 pm/V by resonant PinPoint-PFM method [22].

4.1.1.2.3 Molten Salt

Molten salt synthesis (MSS) is a powerful bottom-up synthesis method of nanomaterials with different chemical compositions and morphologies. This method is preferable in terms of a benign environment, simplicity of operation, low cost, easy scale-up. MSS is mostly commonly employed in processes that require being free of limitations associated with aqueous and organic solvents. In this process, chemical reactions take place in fluxes of salts featured with low melting points. These molten salts eventually act as solvents and aid in efficient diffusion and reaction of the species. MSS differs from flux method in how the latter uses salts as additives while, in the former technique, molten salts act as solvents. The molten salt method of synthesis involves stages of mixing, diffusion, nucleation, and growth in chronological order. Initially, the precursors are mixed with appropriate salt followed by heating at temperatures higher than their melting point to form a molten flux. Thereafter the products nucleate in the molten salt and grow in the desired direction. The precursors are highly reactive in the molten salt, owing to the improved ionic mobility (10^{-5}–10^{-8} cm^2 s^{-1}) and large available contact surface area facilitating a higher reaction rate. In this context, $Pb_{0.95}Sr_{0.05}(Zr_{0.52}Ti_{0.48})O_3$-$Pb(Zn_{1/3}Nb_{2/3})O_3$-$Pb(Mn_{1/3}Sb_{2/3})O_3$ (PZT-PZN-PMS), piezoelectric ceramic was prepared by MSS method and the measured properties were compared to that synthesized by conventional mixed oxide (CMO) route. It was found that the molten salt method improves the sinterability of the ceramics, which results in the improvement of dielectric (ε_r = 1773; tanδ = 0.0040) and piezoelectric (d_{33} = 455 pC/N; k_p = 0.70; Q_m = 888; E_c = 10.3 kV/cm; P_r = 28.2 µC/cm^2) properties than the CMO path [23]. In a similar fashion, nanostructures of crystalline orthorhombic NaNbO$_3$ were synthesized by MSS. Referring to the former method, Na$_2$CO$_3$ and Nb$_2$O$_5$ were added to NaCl as starting materials, which served as molten salt in stoichiometric ratio. The ground mixture was heated to 800 °C and then quenched to room temperature. The samples were then immersed in deionized water for subsequent separation and purification followed by centrifugation and drying. The initial shape and dissolution rate of the precursor determine the final shape of the products. NaNbO$_3$ ceramic nanostructures hold real characteristic of ferroelectric and electric dependence of polarization behavior was found to be greatly influenced by poling treatment. The poled ceramic demonstrated a piezoelectric d_{33} of 28 pC/N which was at par with previous results [24]. Besides,

TMSS (topochemical molten salt synthesis), an extension/modification to this method has the additional advantages of easy control over particle size and shape [25]. This method usually utilizes asymmetric reactants with feeble solubility in required molten salt templates. This is beneficial in controlling the morphology of the designated products through recombination of the local basic unit. Hence, it is a prospective method to synthesize low dimensional symmetrically oriented ferroelectrics, which is otherwise tedious using wet chemical techniques [26, 27]. Zhou et al. [28] employed this approach to prepare $Na_{0.5}Bi_{0.5}TiO_3$-$xSrTiO_3$ (NBT-xST; $x = 0, 0.10, 0.26$) whiskers using $Na_2Ti_6O_{13}$ as the template. At first, the template was synthesized by the molten salt method. Thereafter, it was mixed with the required oxides/carbonate (Na_2CO_3, Bi_2O_3, $SrCO_3$, TiO_2) and salt (NaCl) and, after appropriate steps, the designed whiskers were formed as per the following chemical reactions:

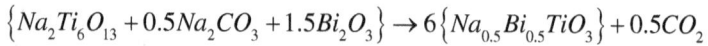

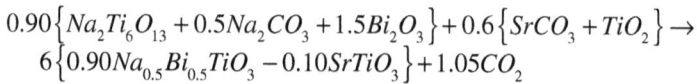

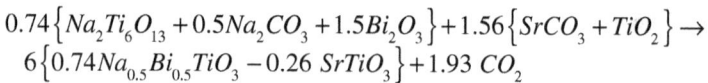

$$\left\{Na_2Ti_6O_{13} + 0.5Na_2CO_3 + 1.5Bi_2O_3\right\} \rightarrow 6\left\{Na_{0.5}Bi_{0.5}TiO_3\right\} + 0.5CO_2$$

$$0.90\left\{Na_2Ti_6O_{13} + 0.5Na_2CO_3 + 1.5Bi_2O_3\right\} + 0.6\left\{SrCO_3 + TiO_2\right\} \rightarrow$$
$$6\left\{0.90Na_{0.5}Bi_{0.5}TiO_3 - 0.10SrTiO_3\right\} + 1.05CO_2$$

$$0.74\left\{Na_2Ti_6O_{13} + 0.5Na_2CO_3 + 1.5Bi_2O_3\right\} + 1.56\left\{SrCO_3 + TiO_2\right\} \rightarrow$$
$$6\left\{0.74Na_{0.5}Bi_{0.5}TiO_3 - 0.26\ SrTiO_3\right\} + 1.93\ CO_2$$

The topochemical transformation from $Na_2Ti_6O_{13}$ into NBT-xST structures was discovered to be the structural rearrangement of the edge sharing octahedral into vertex sharing octahedral. These densely packed whiskers were constructed using $Na_2Ti_6O_{13}$ template possessed favorable aspect ratio with diameter and length in the range 500–800 nm and 5–10 μm respectively. Further, PFM investigations confirmed the existence of piezoelectricity in prepared samples.

4.1.1.2.4 Hydrothermal

Hydrothermal technique deals with heterogeneous reactions for synthesizing inorganic materials above ambient temperature and pressure in an aqueous medium. Generally, the precursors remixed in aqueous media and subjected to heating in a stainless-steel chamber, known as an autoclave, to a temperature above the boiling point of water. As a result, the pressure inside the reaction chamber rises strikingly above the atmospheric pressure. The synergistic effect of high temperature and pressure facilitates a one-step method to fabricate highly crystalline materials eliminating the need of post-annealing heat treatments in most cases [29]. Literature reports the promising piezoelectric properties of lead-free materials using this synthesis methodology. Maeda and his group [30] synthesized (K,Na)NbO_3 ceramics by sintering $KNbO_3$ and $NaNbO_3$ powders prepared by hydrothermal reaction. The densely sintered ceramic exhibited promising electrical properties, including $k_p \approx 0.32$, $k_{33} \approx 0.48$, $Q_m \approx 71$ (radial mode)/118 (33 mode) and a piezoelectric constant $d_{33} \approx 107$ pC/N. In another study, Zhou et al. [31] synthesized NBT with different morphologies – spherical agglomerates of primary nanocubes, nanowires, microcubes with processing temperature in the range 100–180 °C and NaOH concentrations of 2–14M. These morphologies were found to have strong correlations with the synthesis conditions. The authors explain the variation in morphology in terms of in-situ transformation and dissolution-recrystallization. They could make clear observation domain structures and piezoelectric properties using PFM technique. NBT microcubes possessed the highest piezoresponse as compared to spherical aggregates and nanowires. Very recently, Wang and his group [32] studied the effect of hydrothermally synthesized $BiFeO_3$ particles on the microstructure and electrical properties of potassium sodium niobate (KNN) lead-free piezocramics. Optimized piezoelectric properties (d_{33} = 220 pC/N, d_{33}^* = 534.5 pm/V and k_p = 0.46 for $0.993KNN-0.007BiFeO_3$ ceramics. These parameters are higher than the ceramics prepared by solid-state reaction owing to the Schottky barrier generated at the interface of KNNS and $BiFeO_3$. Such a feature facilitates the motion of domain and domain walls leading to superior piezoelectric properties.

4.1.1.2.5 Spray Pyrolysis

Spray pyrolysis is a unique technique of synthesizing nanoparticles or films by creating droplets from a precursor solution, accompanied by evaporation and decomposition in a reactor. A typical high-capacity spray pyrolysis apparatus may have a 1.7 MHz ultrasonic spray generator with a number of vibrators ad reactor and a powder collector. In this process, a mixture of required inorganic compounds is dissolved in appropriate solvent to form the precursor solution, which is atomized into droplets and then transported via a carrier gas to the reactor. Droplets entering the reactor (furnace) become dry and react with the carrier gas to form particles with the limited space of a single droplet. The beauty of this process is each droplet acts as a micro-reactor maintaining the symmetry of the droplet and resulting in spherical particles with no agglomeration. Spray pyrolysis is effective in fabricating particles, which are otherwise difficult to be synthesized with other existing methods, since it involves the formation of metastable state excluding the role thermodynamics. Apart from that, it ensures good distribution of particles at nano scale as the precursor material is precipitated from the micro droplets. This feature is useful in materials at low temperatures with limited risk of contamination [33]. Spray pyrolysis is beneficial for preparing ceramics where coarsening and poor densification is a matter of concern leading to degradation of properties. To avoid such detrimental factors, Haugen et al. [34] employed aqueous spray pyrolysis route to synthesize fine (~100 nm) and phase pure KNN powders with a slim particle size distribution, though they faced some sintering issues. In spite of this hazard, 95 percent dense ceramics with normalized strains up to 180 pm/V could be prepared by normal sintering. In the recent past, nanostructured ZnO films were grown on p-type silicon substrates at different deposition temperatures ranging from 300–500 °C. XRD and PFM confirm a change in the preferred direction of orientation from (002) to (101), with an increase in deposition temperature above 350 °C. A comparison of nanoscale piezoelectric investigations reveals enhanced piezoelectricity when the film deposition was made at 350 °C [35]. In a similar manner, piezoelectric response of ZnO films prepared by ultrasonic spray pyrolysis with preferred orientation along the c-axis was investigated by Ramos-Serrano et al. [36]. In order to evaluate the piezoelectric performance, a prototype based on thin ZnO films deposited on Si cantilevers was constructed. Mechanical deformation was produced by hitting the cantilever with a falling metallic pellet from a certain height. The highest piezoresponse obtained from a single hit was close to 4.5 mV.

4.1.1.2.6   Emulsion Synthesis

The emulsion method is one of the prospective methods of synthesis of ceramic powders, especially multicomponent systems. A solution of aqua-soluble ceramic precursors is emulsified with an organic fluid containing organic surfactant in order to create a dispersion of aqueous droplets of uniform size in the organic fluid. Originally, the aqueous solution is homogeneous, and since the aqueous droplets are uniformly dispersed in the organic phase, each droplet fairly contained the same amount of ceramic powder. Hence, almost spherical and fine powders with minimal agglomerations are produced at relatively low cost. Despite these advantages, there are quite a few reports on the synthesis of ceramic powders by this method. One such investigation was made by Kim et al. [37] in which they synthesized $(1-x)(Bi_{0.5}Na_{0.5})TiO_3-xBaTiO_3$ (x = 0, 0.02, 0.04, 0.06, 0.08 and 0.10) ceramic powders by the emulsion method. The starting materials Bi_2O_3, Na_2CO_3, $Ba(NO_3)_2$ and $TiCl_4$ were dissolved into an acidic medium as per the composition chosen. The mixture was subjected to constant stirring for 24 hours to prepare a homogenous solution. This solution was again stirred to obtain a water-in-oil type emulsion in the presence of span 80 as the surfactant, kerosene as solvent, and paraffin oil as the emulsifying agent. The resulting emulsion was sprayed onto kerosene heated at 170 °C to evaporate water. Ceramic powder was extracted after filtering and drying at 120 °C. The powder was finally calcined and sintered to obtain pellet samples. The samples exhibited superior dielectric and piezoelectric properties with dielectric constant at 1 kHz and piezoelectric constant as 1840 and 174 pC/N respectively.

4.1.2   THIN FILM FABRICATION

A thin film is basically a layer of material with a thickness ranging from a few fractions of nanometers (monolayer) to several micrometers (multilayers). The controlled synthesis of materials in the form of thin films is an extremely essential step in many applications, especially in miniaturized and integrated structures. Properties of thin films depend preferably on their deposition technique. There are various methods for thin film growth (Figure 4.4), a few of them are discussed below.

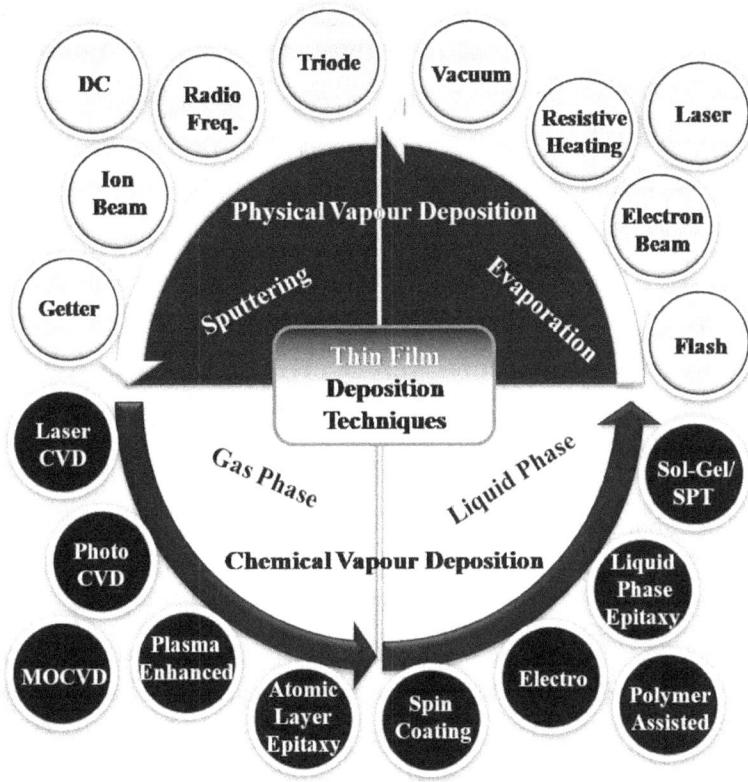

FIGURE 4.4 Overview of various PVD and CVD techniques.

4.1.2.1 Physical Vapor Deposition

Physical vapor deposition (PVD) mainly relies on the purging of atoms from solids or liquids by the application of energy followed by deposition on a substrate material. It covers a range of film deposition techniques by physical means, including evaporation, vacuum arc, laser ablation, sputtering, and so forth. All these processes generically involve individual atoms or groups of atoms, which may not be found in the gaseous phase. Usually, these atoms are scratched from a solid or a liquid, forced to travel via an evacuated chamber and, thereafter, subjected to impingement on the substrate. On the surface of the substrate, these atoms attach themselves and form a thin layer called *film*. The atom-purging process may be thermally assisted, as in evaporation and ablation or by sequential collision of energetic particles like electrons, ions, and so forth, as in a sputtering deposition [38]. An illustration portraying the basic principle of PVD and; comparison among the most common PVD techniques, that is, sputtering and evaporation, is displayed in Figure 4.5.

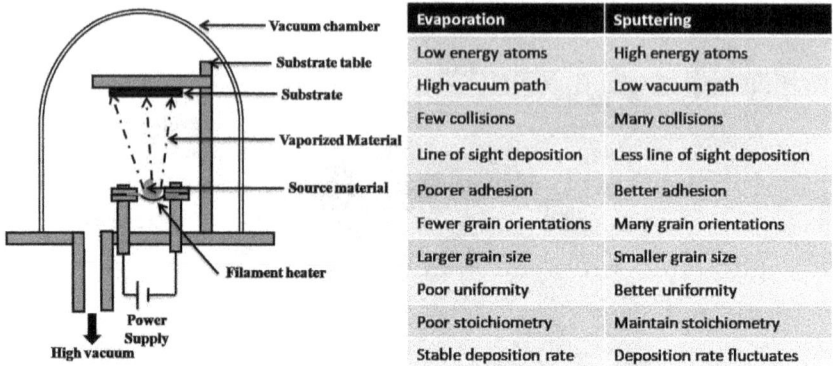

Evaporation	Sputtering
Low energy atoms	High energy atoms
High vacuum path	Low vacuum path
Few collisions	Many collisions
Line of sight deposition	Less line of sight deposition
Poorer adhesion	Better adhesion
Fewer grain orientations	Many grain orientations
Larger grain size	Smaller grain size
Poor uniformity	Better uniformity
Poor stoichiometry	Maintain stoichiometry
Stable deposition rate	Deposition rate fluctuates

FIGURE 4.5 Basic principle of PVD (left) and comparison between evaporation and sputtering (right).

4.1.2.1.1 Sputtering

As discussed above, the majority of the applications of sputtering involve ion/ neutral atom bombardment on the target surface generating large fluxes of energy. Based on this principle, the deposition systems may be plasma or ion-beam based. In the case of plasma systems, ions from the plasma beam are usually bombarded on a cathode, which gets sputtered off under appropriate temperature and pressure conditions and deposited onto the nearby target surface. The sources of plasma may be diode; DC and RF diodes or magnetron. Sputtering using RF magnetrons have been the interest of researchers for many years [39, 40]. In 2010, Ababneh et al. [41], sputter-deposited AlN thin films from Al target with an interesting piezoelectric property and high Complementary Metal Oxide Semiconductor (CMOS) capability. Since a good c-axis orientation is highly essential for achieving superior piezoelectric properties, under such circumstances, sputtering conditions were given utmost priority. It was found that highly c-axis oriented films could be deposited with nominally unheated (100) Si substrates. The degree of orientation increased with lowering sputtering pressure and increasing N_2 concentration. The piezoelectric coefficients of c-axis oriented 500 nm films determined experimentally by laser scanning vibrometry were found to be $d_{33} \approx 3.0$ pm/V and $d_{31} \approx -1.0$ pm/V. Similarly, InN thin films were deposited on Si(100) and Pt(111)/Ti/SiO$_2$/Si(100) substrates by reactive magnetron sputtering. With a plasma power of 25 W, nitrogen gas pressure of 5–7 mTorr and substrate temperature of 300–400 °C, (0002) orientation of the film was achieved. The reason for this preferential growth may be stated as: at lower values of plasma power, the growth rate of the film is slow owing to less ion concentration. Under such circumstances, atoms have enough time to migrate on the growth surface and preferably grow in a particular direction (0002). Growth along a favorable direction is responsible for inducing piezoelectric property in the thin films and, hence, in this case, piezoelectric coefficient measured by heterodyne interferometer was found to be 3.12±0.10 pm/V [42].

Further, in case of diode plasma, the target is generally exposed to plasma, which may cause undesirable heating and other plasma-based damage. An alternative to this is the use of an ion beam source (known as thrusters). A beam of ions generated from the source are directed towards the target from which atoms are sputtered onto the sample surface. In this regard, unusual $(10\overline{1}0)$ orientation of ZnO films was observed, in which the c-axis lies in the substrate plane when irradiated by the ion beam. This in-plane textured $(10\overline{1}0)$ film deposited via ion beam irradiation (0.5–1 keV) excited a shear acoustic mode without any longitudinal wave. An appreciable shear-mode electrochemical coupling coefficient of 0.16 was obtained using the film irradiated with a power of 1 keV. This value is almost 60 percent of that of ZnO single crystal [43].

4.1.2.1.2 Evaporation

Evaporation is a common method of thin film deposition in which the material to be deposited is evaporated in a vacuum. The presence of vacuum guides the path of the vapor particles towards the target or substrate, where they condense back into solid state. Evaporation sources may be categorized as quasi-equilibrium and non-equilibrium sources, both having a wide range of applications. In the quasi-equilibrium category, evaporation occurs close to equilibrium with its vapor. One of the examples of such sources can be an effusion or Knudson cell, which is usually a closed chamber with a small aperture – smaller than the remaining internal surface area. In such a case, evaporation losses follow a perturbation on the dynamics of liquid-vapor equilibrium in the chamber. On the other hand, non-equilibrium sources are open sources in which a small area of a material in liquid phase evaporates into a larger, low-pressure volume. Due to the low pressure of the material in a liquid state, this is said to a non-equilibrium source. Common examples of such sources are boat, crucible and e-beam sources. The boat is usually made out of a refractory material such as W or Ta, and is heated by the passage of a large current through a band of metal forming the boat. Besides, a crucible is usually a ceramic cup wrapped within a coil of metallic wire heated resistively by passing a huge current. On the other hand, the most widely used source, e-beam, employs an electron beam originating from a filament placed underneath or adjacent to the evaporation point. Several works have been performed using the evaporation techniques (with different sources) to deposit piezoelectric thin films on desired target material. In this context, highly oriented single-phase ZnO thin films were grown by a simple e-beam evaporation method at different substrate temperatures and at room temperature followed by subsequent annealing in an oxygen atmosphere at different temperatures. The optical properties of the films can be tuned by increasing the substrate temperature or a post-deposition annealing temperature. It was also observed that the optical and electrical properties also improved by annealing of the film in an oxygen ambient after deposition [44]. Similarly, Periasamy et al. [45] had successfully grown ZnO nanoneedle arrays on Si substrate using the thermal evaporation technique. Prior to the thin film growth, 20 mm thick layer of Al doped ZnO with ~1 percent Al was deposited on Si wafer as a seed layer via thermal evaporation. This buffer seed layer favors the further growth of ZnO nanoneedles and controls the orientation of the film during growth. The rectifying platinum contact on ZnO nanoneedle arrays were able to efficiently convert nanoscale mechanical energy into electrical energy by exploiting the piezoelectric and semiconducting properties of ZnO.

4.1.2.2 Chemical Vapor Deposition

Chemical vapor deposition (CVD) is a very powerful technique for fabricating fine quality and high-performance solid thin films and coatings. In this process, thin films are created on a heated substrate surface with the help of a chemical reaction of gas phase precursors. Formation via chemical reactions is utmost beneficial as the deposition rates can be tuned to obtain high quality films with superior conformality. With advancement in time, several variants have been added to the basic CVD technique, such as low-pressure CVD (LPCVD), plasma enhanced CVD (PECVD) and metal-organic CVD (MOCVD). However, in all types of CVD, some common elementary steps are followed. In the first step, the reactant gases are introduced to the reaction chamber. In the reactor, the initial reactants undergo gas-phase reactions to form some intermediate reactants and by-products in gaseous phase through homogeneous reactions or diffuse directly to the substrate through the boundary layer. In either case, the initial gaseous and intermediate reactants adsorb onto the heated substrate and diffuse on the surface. Adsorption is followed by heterogeneous reactions at the gas–solid interface resulting in thin film formation via nucleation, growth, and coalescence. These heterogeneous reactions also lead to the formation of reaction by-products. In the final step, the unreacted species and gaseous by-products are removed from the reaction zone via desorption. Gas phase reactions usually happen when the temperature is extremely high or some additional energy is added, for example, plasma phase. Further, heterogeneous reactions become essential at places where the deposition reaction relies on the surface catalysis of the underlying substrate [46]. The basic principle and instrumentation of LPCVD, MOCVD and PECVD are presented in Figure 4.6.

4.1.2.2.1 Low Pressure CVD

Low pressure CVD refers to a process in which chemical reactions occur under a low working pressure, which is maintained with the help of a vacuum pump. It not only decreases the formation of any unwanted gas phase reactions, but also increases the uniformity across the substrate. Preparation of piezoelectric thin films using this method is rarely found in literature. A preferred methodology for their synthesis is metal-organic CVD method.

4.1.2.2.2 Metal-Organic CVD

Metal-organic CVD or MOCVD is one of the advanced versions of chemical vapor deposition, which employs metal organic precursors. The fundamental

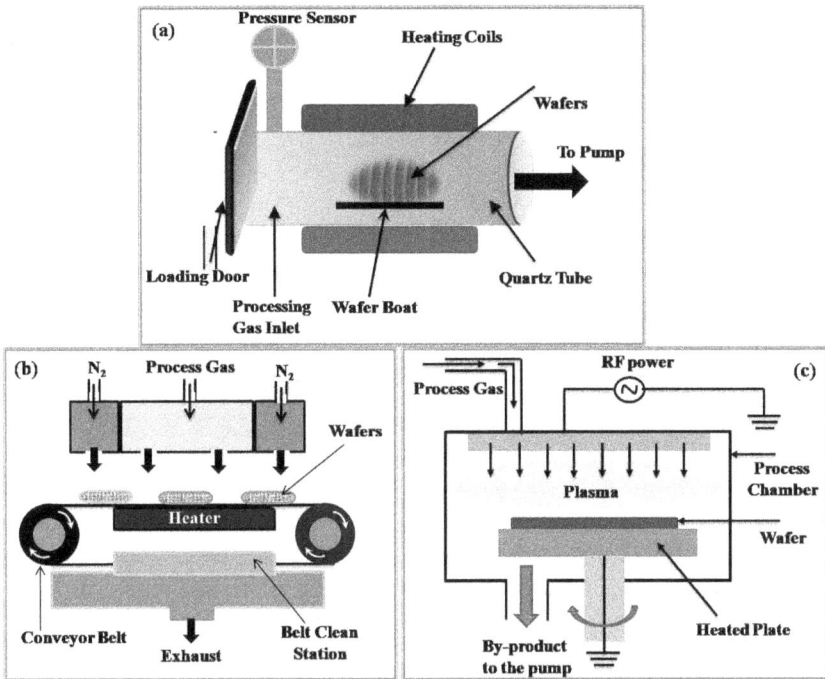

FIGURE 4.6 Basic principle and instrumentation of (a) Low pressure CVD; (b) Metal organic CVD and (c) Plasma enhanced CVD.

principle of MOCVD is to achieve contact between volatile precursor materials and the heated substrate surface. The precursors are initially directed towards the substrate by some carrier gas (Ar/N) and then absorbed on its surface. The reactive species thus diffuse at the sample surface to their preferential site and react in a heterogeneous phase, giving rise to the formation of films. The volatile by-products left behind by CVD are discharged to the glass flow vectors. In 2021, Kim et al. [47] used GaN thin films deposited by the MOCVD method to fabricate a flexible eye movement sensor for sensing the blinking of the human eye and eyeball motion as a part of personal healthcare, safety, and entertainment systems. For the synthesis of GaN films, III-N layers were epitaxially grown by a metal-organic chemical vapor deposition on (111) Si substrate using trimethyl ammonium, trimethylgallium and ammonia as the gaseous precursors and hydrogen as the carrier gas. Initially an AlN buffer layer was grown on Si substrate with a thickness of about 100 nm with no interactions between the buffer layer and the substrate. Thereafter, graded aluminum gallium nitride layers were grown with increasing gallium nitride mole fractions up to a total thickness of 600 nm. This piezoelectric film sandwiched between PDMS layers acted as good mechanical sensors to detect the various motions of the eyelid and eyeball.

Previously, GaN thin films were grown on (111) Si substrate with AlGaN buffer layer using a conventional shower head type metal or organic chemical vapor deposition reactor. The films demonstrated prominent electro-optic and converse piezoelectric properties [48]. Similarly, MOCVD has also been applied to deposit other types of thin films, such as $BiFeO_3$ [49].

4.1.2.2.3 *Plasma Enhanced CVD*

Plasma enhanced CVD or PECVD is a very useful technique to deposit thin films of various materials on a target at lower temperatures than the conventional CVD method. It is a hybrid coating process in which chemical vapor deposition is activated by energetic electron (100–300 eV) within the plasma as against the thermal energy associated with standard CVD processes. Many researchers have done work in this direction, a few of whom are quoted here. Abdallah et al. [50] deposited polycrystalline aluminum nitride films using a microwave plasma enhanced CVD at different deposition temperatures. Well crystallized <0001> orientation was obtained at a low temperature of 500 °C. However, the sample exhibited better piezoelectric performance for the film deposited at 700 °C demonstrating a film of thickness 1 μm and d_{33} of 5.8±1.2 pm/V. Although efforts are made to achieve good crystallinity of the films, it is often observed that the deposited films have inferior piezoelectric properties due to random orientation of the crystal structure. Since low crystallinity has adverse effects on the piezoelectric property of the film, many modifications have been suggested in recent years. Goff et al. [51] exploited the benefits of combining DC magnetron sputtering and radio frequency-plasma enhanced CVD to improvise the film properties. Though the films prepared by them could not be fully characterized due to preliminary nature of the process, it was noticed that a uniform film deposition was achieved at a faster rate. Al doped ZnO films were deposited via remote plasma enhanced metal-organic CVD using oxygen/ diethylzinc/ trimethylaluminium mixtures to understand how the material properties such as grain size, and morphology are affected by working pressure. This is because working pressure plays a key role in controlling the sheet resistance, which is otherwise detrimental for application purposes [52].

4.1.2.3 Chemical Solvent Deposition

Chemical solvent deposition (CSD) is a very comprehensive term, which can be employed to describe any technique that refers to the use of a chemical precursor solution to create a film. It is a kind of soft solution used to create a film through a simple experimental set up under ambient temperature and pressure. In this technique, no vacuum processing is required. Though CSD can be activated by hydrothermal and electrochemical reactions, simple CSD applying high temperatures or electrochemical power is much more advantageous, since it can be used for low heat resistant materials, porous materials, and also complicated shapes. Some of the important chemical solvent deposition techniques for thin films are discussed below.

4.1.2.3.1   Sol-Gel

Sol-gel films are usually formed by gravitational or centrifugal draining accompanied by drying. The nature of the films largely depends upon the timescale of deposition, shape of the liquid profile, and forces exerted on the solid phase. In this process, sol (precursor solution) is prepared in the same way as discussed in section 4.1.2.2. The sol, or the precursor solution, is then subjected to dip coating, spin coating, or drain coating to coat the substrate surface, followed by drying. In dip casting the substrate is pulled out of the sol bath at a constant speed due to which the solution is entrained in fluid mechanical boundary. The boundary layer separates into two above the sol bath, from which the outer layer is restored into the liquid. Due to continuous evaporation and drainage of the solvent, a probable wedge-shaped thin film is formed. While drying, if the dry line velocity comes close to withdrawal speed, a steady stable is achieved with respect to liquid bath surface leading to uniformity in the thin film. On the other hand, spin coating differs from dip coating in the way the deposited films are dried by centrifugal draining and evaporation. Dutta et al. [53] deposited ZnO thin films by the sol-gel drain coating technique for varying concentration (0.03, 0.05, 0.08 and 0.1 M) of sol, which is prepared with continuous stirring of zinc acetate 2-hydrate of different concentration in isopropyl alcohol medium. Simultaneously, Diethanolamine was added to the mixture as sol stabilizer. After preparation of the sol, the glass substrate was drain coated with it at a drawing speed of 4 cm/min by the sol. In between each coating, the substrate was dried at 350 °C for 10 minutes. In the final stage, the substrates were air heated at 550 °C for 1 hour to obtain ZnO films. These films demonstrated a high degree of crystallinity and optimum electrical properties. Multicomponent piezoelectric systems such as $(Bi_{0.5}Na_{0.5})_{0.95}Ba_{0.05}TiO_3$ and Mn doped $(Bi_{0.5}Na_{0.5})TiO_3$-$(Bi_{0.5}K_{0.5})TiO_3$-$SrTiO_3$ thin films were also synthesized on $Pt/TiO_2/SiO_2/Si$ substrates via the sol-gel processing technique [54, 55]. Both films exhibited superior piezoelectric properties as estimated by PFM technique.

4.1.2.3.2   Electrochemical Reaction

Electrochemical deposition or electro deposition is one of the novel techniques that include the processing of a wide range of materials, including metals, polymers, and ceramics. This process facilitates the effective deposition of films with the aid of external electric potential. Hence the substrate surface must have an electrically conducting coating before deposition takes place. Optimized coatings can be obtained with well tuned deposition parameters, liquid bath composition, anode and cathode material. Some researchers have done work in this direction. In 1990, Yoo et al. [56] deposited polycrystalline strontium titanate thin films on a titanium target by a hydrothermal-electrochemical method. In the electrochemical cell, thoroughly cleaned Ti plate was used both as the anode or working electrode and cathode or counter electrode along with $Sr(OH)_2$ as the electrolyte. This electrochemical cell was placed in an autoclave and heated from ambient temperature to 200 °C. The resulting films exhibited smooth and homogeneous surfaces with limited porosity. Piezoelectric $BaTiO_3$ thin films were electrochemically coated on a titanium target by Lu et al. [57], using highly alkaline $Ba(CH_3COO)_2$ and NaOH electrolyte. They observed that the TiO_2 rutile phase was formed at a constant current of 10 mA while the $BaTiO_3$ cubic phase was formed at currents above 20 mA. Variation in time at 30 mA also leads to the formation of the rutile phase in 30 minutes and barium titanate after 45 minutes. This shows that a change in the current and time of deposition can affect the formation of the films. Thin films with uniformly distributed small grains owing to dissolution and crystallization could be achieved at lower electrolytic currents and shorter deposition times. In yet another interesting report, ZnO nanorods were grown by a template-free electrochemical deposition approach, and its piezoelectric properties were assessed. Large arrays of ZnO nanorods vertically aligned to the substrate demonstrated high aspect ratio with a well-defined hexagonal symmetry and high crystallinity. The effective piezoelectric coefficient extracted from piezo force microscopy yield a value as high as 11.8 pm/V. the combined results advocates the application of piezoelectric devices based on ZnO nanorods with superior performance [58].

4.1.2.3.3   Hydrothermal

Hydrothermal method is a unique technique of thin film deposition which offers the possibility of coating a range of ceramic thin films at low temperatures (90–200 °C). It not only eliminates the need of high-temperature calcination steps but also generates homogeneous thin films with uniform grain size and crystallinity. The nucleation and growth of the films can be tuned by adjusting the reaction parameters such as temperature, pressure, and mineralizer / additive concentration. In this regard, the effect of reaction time and concentration of mineralizer were studied on the grain size, crystallinity, surface roughness, and thickness of barium titanate thin films. The experimental procedure includes the preparation of a feedstock out of $BaCl_2.2H_2O$ or $Ba(OH)_2.8H_2O$ along with KOH mineralizer. The mineralizer stabilizes the temperature and lowers it for crystal formation. The feedstock along with Ti deposited Si(100) wafer were transferred to an autoclave and heated at 140 °C for the reaction to occur. The concentration of the mineralizer and reaction time were varied and, after cooling the apparatus, the final product was obtained, which was cleaned thoroughly to remove the impurities. It was observed that the grain size reduced with increase in KOH concentration due to the rise in supersaturation of the reactant solution but, on the other hand, surface roughness increased. Hence, an optimized amount of mineralizer with appropriate reaction time was necessary to grow good-quality thin films [59]. In yet another work, $BaTiO_3$ thin films patterned as nanotubes with honeycomb morphology could be grown on Ti substrates in a two-step modified hydrothermal process. In the first step, titanium substrate is anodized to generate a surface layer of amorphous TiO_2 nanotube arrays. Second step involves the hydrothermal treatment of anodized substrates in aqueous $Ba(OH)_2$ in which the nanotube arrays act as the template for conversion into polycrystalline barium titanate nanotubes [60]. A similar two-step hydrothermal process was also employed to deposit PZT films by controlling the rate of nucleation and growth. The first step was a nucleation step in which TiO_2 substrate reacted with a combination solution of Pb and Zr to form PZT/PZ nuclei on its surface. Further, the crystal growth of PZT was promoted in the second step, which was a hydrothermal reaction (crystal growth process) of Pb, Zr and Ti. Though the films exhibited good electrical properties, the process needed modification to remove defects and minimize the dielectric loss [61].

4.1.2.4   Chemical Melt Deposition (Liquid Phase Epitaxy)

Epitaxy refers to a class of crystal growth involving deposition of new, or more crystalline, layers with well-defined orientations with respect to the crystalline seed layer. Such a crystalline film is called an epitaxial film. When the crystal is grown from melt on solid substrates, it is known as liquid-phase epitaxy. In this method, growth is carried out from the precursor solution in a relatively simple and inexpensive apparatus at low temperatures under near equilibrium conditions. This results in films of high purity and a minimum density of defects at low operational cost. In this method, deposition can be performed either from diluted solutions at low temperatures or using concentrated solution at high temperatures and even from melts close to the melting point. However, epitaxy is preferred mostly from dilute solutions, since it allows for control over the thickness of the film due to a slow growth rate. In addition, growth at lower temperatures supports structural perfection and stoichiometry, reduces the adverse effects of differences in thermal expansion coefficients between the substrate and film, and impedes the nucleation of spontaneously nucleated crystallites [62]. This technique is usually employed to design thin films with optical properties and is rarely reported for piezoelectric thin films. One of the works was done very recently by Zhang et al. [63] in which they had grown piezoelectric α-quartz films on silicon (100) substrate. They observed that film thickness, relative humidity, and nature of the surfactant play important roles in controlling the microstructure, homogeneity, and crystallinity of the films.

4.1.3   SINGLE CRYSTALS

Single crystals are known for uniform, continuous, and highly ordered structures that result in unique and exotic properties as compared to polycrystalline substances. Owing to their superior properties, single crystals have been in extensive use in optical, electronic, piezoelectric and other applications. Especially, piezoelectric single crystals are technologically important for sonars, medical diagnostic devices, energy harvesters, and many other applications. Generally, the single crystals are grown by three types of approaches: from melt, solution, and vapor phase.

4.1.3.1   Melt Growth Techniques

Melt growth is one of the most commonly used techniques based on the solidification and crystallization of a melted material. Crystal growth from melt offers the advantage of growing large single crystals of superior quality in a relatively short time as compared to other methods. In spite of these advantages, it is also associated with some difficulties in maintaining a stable environment, attaining a very high melting point of materials, preserving chemical homogeneity in the case of multi-component systems, and involvement of high temperatures. Czochralski and Bridgeman are the most widely used melt growth techniques. Apart from those, there are a few other melt growth techniques such as Vernuil, zone melting, Kyropoulos, and so forth.

4.1.3.1.1 Bridgeman

Crystal growth via the Bridgeman method refers to the directional solidification of a liquid (melt). In this method, the molten material is slowly cooled by moving its container from the hot to the cold zone. In order to begin the process of crystal growth, the end of the container is elongated to place the seed crystal. The incorporation of the seed crystal requires precise temperature control at the interface, though in some cases, growth takes place without the seed. In the conventional Bridgeman method, cold zone is placed outside the furnace and, hence, it is difficult to maintain the temperature gradient. However, in the modified Bridgeman–Stockbarger method, two finely controlled temperature zones are made by introducing two separate furnaces with a baffle in between. Some other variations in the technique involve rotating containers and horizontal arrangement of furnaces(s). Important parameters that affect the growth are container material, hot zone temperature, temperature gradient, cooling rate and so forth. Xu et al. [64] synthesized [011] oriented PMN-xPT single crystals by modified Bridgeman method and studied the domain structure evolutions during the poling process. Room temperature poling suggested that structural transformation sequence firmly depends on the composition. It was found that a direct transition to 2R/2T domain state in rhombohedral or tetragonal phase field beyond the morphotropic phase boundary (MPB) while in the MPB region, it was difficult to achieve the said domain state, though the initial state remains either rhombohedral or tetragonal. Around the morphotropic phase boundary, PMN-PT exhibited weak piezo-properties (d_{33} = 250 pC/N) while an extraordinarily high piezoelectric constant, d_{33} = 1000 pC/N was obtained in [011] oriented multidomain ceramics. Hence, the modified Bridgeman method was effective in the growth of piezoelectric [011] oriented single crystals and may be extended to crystal orientations like [001] and [111].

4.1.3.1.2 Czochralski

Czochralski process is one of the most important techniques for the fabrication of bulk single crystals for a wide range of applications. In this method, initially the feedstock is kept in a cylindrical crucible and melted by resistance or radio-frequency heaters. Once the feedstock melts completely, a seed crystal having a typical diameter in the range of few millimeters is dipped from the top into the free melt surface due to which a small portion of the dipped seed is melted. At the interface of the seed and the melt, a meniscus is formed. Thereafter the seed is drawn back from the melt at a slow pace (often under rotation) and crystallization of the melt takes place resulting in a new crystal portion. During subsequent growth, shape of the crystal, especially the diameter is regulated by carefully adjusting the heating power, pulling rate and rate of rotation of the crystal. The diameter of the crystal can be controlled in this method either by controlling the meniscus shape or weighing the crystal/ melt. In the past, high quality $La_3Ga_{5.5}Nb_{0.5}O_{14}$ piezoelectric single crystals having applications in bulk and surface acoustics were grown by the conventional Czocharlski process. The starting materials (La_2O_3, Nb_2O_5 and Ga_2O_3) were mixed in stoichiometric amounts and pressed into tablets before sintering to ensure crystal growth from the polycrystalline compound. Single crystals were then grown via RF-heating Czochralski technique in a TDL-J40 single crystal growth furnace. Once the growth was complete, the crystal was brought down to ambient temperature at cooling rate of 15–50 °C/hour. High resolution XRD confirmed the formation of high-quality crystals yielding superior piezoelectric behavior [65].

4.3.1.3 Verneuil

The Verneuil method, also known as flame fusion, is a process initially used to manufacture gemstones. However, eventually it was applied for preparing a variety of oxides. Since no crucible is required in this technique, issues related to contamination or reaction with crucible material is avoided. Unlike other crucible-less methods such as skull fusion, float zone fusion, electron beam fusion, and so forth, flame fusion can be employed to synthesize materials that are non-conductors of electricity. The basic methodology involves the melting of a finely powdered substance in an oxyhydrogen flame and crystallizing melted droplets into a boule. The starting materials are taken in the form of fine powder. This powder is sprinkled from a supply container into the oxyhydrogen flame, which reaches the melt film formed on top of seed crystal. An amount of matter corresponding to the powder flow crystallizes continuously out of the film, and the crystal is pulled downward at the same rate as it is growing towards the flame. Radiation heat losses are minimized by enclosing the crystal in a chamber. As soon as the crystal grows to a desired size, powder flow and crystal motion are ceased; and the flame is extinguished. The use of this technique for growing piezoelectric single crystals is rare in the literature. However, Mancini et al. [66] synthesized Mn doped $SrTiO_3$ single crystals using Verneuil growth process. The crystals exhibited optimum structural and electronic properties.

4.1.3.2 Solution Growth

Crystallization of fairly soluble compounds from supersaturated solutions under ambient pressure conditions inappropriate non-reactive solvents is generally referred to crystal growth from solution. The most commonly chosen solvents for this method are different organic solvents, water and/or their mixtures, and the melt of any chemical compound. Organic liquids and water are liquids under ambient conditions, while some of the chemical compounds take the form of liquids at elevated temperatures. Based on the growth temperature, which depends on the existence of the solvent in liquid state, this technique can be differentiated as high and low temperature solution growth. Other solution growth techniques include hydrothermal and gel growth.

4.1.3.2.1 Hydrothermal

Hydrothermal single crystal growth offers a suitable alternative to the conventionally used crystal growth methods to fabricate new materials and crystals for specific applications. This method is capable of growing crystals well below their melting points and can potentially produce bulk crystals with less reduced thermal strain. Hydrothermal crystal growth is quite similar to the growth of crystals from aqueous solutions at room temperature. However in this process, crystal growth takes place in a closed chamber (autoclave) and follows a proper regime of pressure and temperature. The temperature gradient maintained in the autoclave facilitates the transport of solute particles from the hot region to the cold region via convection. In a recent report, hydrothermal process was employed to grow microcrystalline $GaAsO_4$ powder in an oxidizing environment (presence of H_2SO_4), which later acted as a nutrient for developing $GaAsO_4$ single crystals by epitaxy. In order to grow large single crystals of $GaAsO_4$, a specialized autoclave comprising two different temperature regions was used. One of the zones acted as the dissolution zone in which the nutrient was kept at the dissolution temperature while seed crystals were placed in the second zone where the crystal growth takes place at a higher temperature. This temperature gradient across the two zones results in the diffusion of $GaAsO_4$ from dissolution to crystallization region [67]. Prior to this, Cambon et al. [68] $GaAsO_4$ single crystals were also produced by using a single PTFE-lined autoclave to obtain large single crystals.

4.1.3.2.2 Gel Growth

Crystal growth via gel medium is one of the simplest and well-suited techniques for compounds that are sparingly soluble in water and decompose before melting. The experimental requirement for this method can be set up in the laboratory with simple glassware, without the need of any sophisticated instruments. Usually, it is known that gel is a semi-solid type two component system which is rich in liquid, stable in form, flexible, and has fine pores to facilitate diffusion through them. The gel can be of many types but silica hydrogel or agar is the most liked one for the crystal growth process. Initially, a gel solution of suitable density is prepared and impregnated at a proper pH by one of the reactants and then transferred to a test tube. After some time when the gel sets, a supernatant solution (appropriate molar concentration) is poured slowly on it. This solution now diffuses into the gel and reacts with it. Thereafter, nucleation and growth begin and this process is found very similar to the growth of a fetus in the womb. In spite of all advantages, the growth of piezoelectric crystals by this method is rare in the literature. It is associated with some inherent disadvantages, like the size of the grown crystals is small but the period of growth is long [69].

4.1.3.2.3 Low Temperature Crystal Growth

Low-temperature processes involve the growth of single crystals at or below ambient temperature. It can be of three types – slow cooling, solvent evaporation, and temperature gradient. In the slow cooling method, a saturated solution at temperatures above room temperature is kept in a crystallizer, and a seed crystal is suspended in the solution. The crystallizer is placed in a water thermostat, which is programmed to cool as per requirements. Besides, in the solvent evaporation method, an extra amount of solute is entrenched by using the gradient between the rate of heating and cooling. In the process of entrenchment, there is loss of volume, since weakly bound particles are lost. This is in contrast to the slow cooling method in which the volume of the solution remains constant. Generally, in the solvent evaporation method, the vapor pressure of the solvent remains higher than the solute due to which solvent evaporates more rapidly, making it supersaturated. The third process, called the temperature gradient method, involves the transport of material across a temperature gradient that is, from a hot source to the cooler supersaturated solution resulting in crystal growth.

4.1.3.2.4 High Temperature Crystal Growth

High temperature crystal growth, also known as flux growth is an important technique of growing single crystals for a wide range of materials, in particular multicomponent systems. This methodology is beneficial since the crystals are grown below their melting point. However, in case of materials decomposing or exhibiting phase transitions below the melting point and/or having a high vapor pressure at the melting point, the crystal growth has to be carried out at temperatures lower than phase transition temperature. Besides, the main disadvantage of this technique is the possible existence of a large pool of impurities in the grown crystals, either as solvent inclusions or substitutions. That is the reason, it is difficult to achieve desired levels of purity, though crystals with finer perfection can be developed [70]. Further, supersaturation of the solvent and crystal growth in this method are obtained majorly by slow cooling, solvent evaporation or temperature gradient transport, as mentioned in the earlier section 4.3.2.3. Some of the piezoelectric single crystals (both lead-based and lead-free) have been fabricated in the past. $[Bi_{0.5}(Na_{1-x}K_x)_{0.5}]TiO_3$ single crystals of size 10 x 10 x 5 mm³ were grown successfully using the self-flux method. Synthesis of larger crystals was facilitated by controlling and restraining multisite nucleation and was achieved via top-cooled solution growth. Along with good dielectric property, the crystals demonstrated a high Curie temperature ~365 °C and a piezoelectric constant of 160 pC/N along [001] direction [71]. Similarly, Singh et al. [72] developed $Pb[(Zn_{1/3}Nb_{2/3})_{0.91}Ti_{0.09}]O_3$ single crystals by flux growth process with some modifications in the temperature profile to enhance the amount of perovskite phase formation and suppress pyrochlore phase formation.

The system exhibited optimum dielectric and piezoelectric properties due to suppression of the pyrocholore phase.

4.1.3.3   Vapor Growth

Single crystals can be grown from the vapor phase by sublimation provided the material has sufficiently high vapor pressure at the processing temperature. This method is mainly employed for the fabrication of semiconductor crystals rather than piezoelectric. In this process, specific chemical precursors containing the desired elements (vapor form) are introduced into a reactor using an appropriate gas carrier. These precursors combine with each other to form the nutrient phase before reaching the substrate material. Release of elements necessary for growing crystals occurs at the solid-gas interface or directly in the gaseous phase, depending on the thermodynamics involved. Vapor growth technique results in low dislocation and defect density as compared to melt/solution grown crystals. In addition to that, this method is most suitable for materials that melt incongruently. It basically involves three steps – vaporization, transport and deposition. The vapor is created by heating the required solid or liquid to high temperatures. Next the vapor is transported through the vacuum driven by the kinetic energy of vaporization. Finally the deposition of vapor occurs via condensation or chemical reaction. Vapor phase growth methods can be distinguished on the basis of the source used and the mechanism of growth. The simplest of all is sublimation, in which precursor material is placed at one end of a sealed tube and heated to initialize sublimation. The sublimated material is then transported to the cooler region of the tube to enable crystallization. Other vapor phase methods include metallo-organic epitaxy, plasma-assisted epitaxy, hybrid epitaxy, molecular beam epitaxy, and so forth [73].

4.1.3.4   Solid State Crystal Growth

The most convenient way of developing piezoelectric single crystals, which has attracted the attention of the research community, is solid-state conversion of polycrystals into single crystals. This is based on a feature usually observed in many complex systems while sintering, and known as abnormal grain growth (AGG). It is cost-effective as well as being a simple way of obtaining single crystals that can support mass production for various applications. Apart from that, it lessens the number of machining steps after the growth process, which is usually associated with conventional techniques allowing the growth of crystals in more complex shapes. In most of the practical cases, solid-state crystal growth (SSCG) utilizes a seed of similar crystalline structure as the matrix material, either embedded in the green body of the polycrystal or placed on top of it (seeding method). Thereafter, both the seed and the green body are subjected to sintering at a temperature lower than the melting point of the material. This results in the conversion of polycrystal into a single crystal through a controlled abnormal grain coarsening (AGC) process. Besides, the said process can also be seed-free on applying a temperature gradient and/or adding a proper dopant. So far, SSCG has been successfully applied to a number of solid oxides and piezoelectrics [74]. Kang and his group have done appreciable work in growing piezoelectric single crystals. They fabricated high quality $95Na_{1/2}Bi_{1/2}TiO_3$-$5BaTiO_3$ (NBT-5BT) single crystals of size 10 mm x 10 mm x 3 mm by the solid-state single-crystal growth method (Figure 4.7). The abnormal grain coarsening behavior of the same NBT-5BT system was also studied. The samples sintered at 1200 °C exhibited abnormal grain growth with increase in sintering time. The crystal growth mechanism was mainly governed by the growth of facet planes with a low energy step [75]. Initially, NBT-5BT single powder was synthesized by solid state sintering technique using commercial grade powders (110). $SrTiO_3$ was used as the seed crystal and it was embedded in single-phase calcined NBT-5BT powder. Powder along with the seed was lightly isostatically pressed into disks at 200 MPa and then sintered in air on a Pt plate in an alumina crucible at 1200 °C for a suitable time. The developed single crystals were then separated from the seed crystal and remaining polycrystal by a low speed cutter. XRD and Raman spectroscopic studies revealed rhombohedral single phase crystals. The [001] oriented single crystals (Figure 4.7) demonstrated excellent dielectric and piezoelectric properties with dielectric constant ~723, loss factor ~0.01 and piezoelectric constant ~207 pC/N and coupling coefficient ~0.5. These reported figures were superior to melt-grown crystals [76]. In another work, incorporation of TiO_2 also had a positive effect on abnormal grain growth behavior. Doping with TiO_2, changed the grain shape towards more faceted and abnormal indicating an increase in step free energy, thereby raising the critical driving force for appreciable growth. Moreover, it leads a reduction in the number of grains having driving forces greater than respective critical driving force for considerable growth. This was the suggested mechanism for abnormal grain growth with increase in TiO_2 addition [77] and a prospective candidate of single crystal growth.

4.1.4 COMPOSITES

Engineering composites with complex structures for piezoelectric applications need a very systematic and interactive approach to achieve optimum material properties. Composites are basically multicomponent systems with two or more constituent elements resulting in properties that might be different from individual constituents. Piezoelectric composites possess improved electrical properties depending on the synergistic effect of the components. There are various techniques for the fabrication of composite structures; a few important ones are discussed here.

4.1.4.1 High Temperature Dielectrophoresis Technique

Dielectrophoresis (Figure 4.8(a)) is the phenomenon of generation of force by a dielectric material when it is subjected to a nonuniform electric field. Most importantly, the material does not need to be charged; rather all materials can display dielectrophoretic activity in the presence of electric fields. Often it

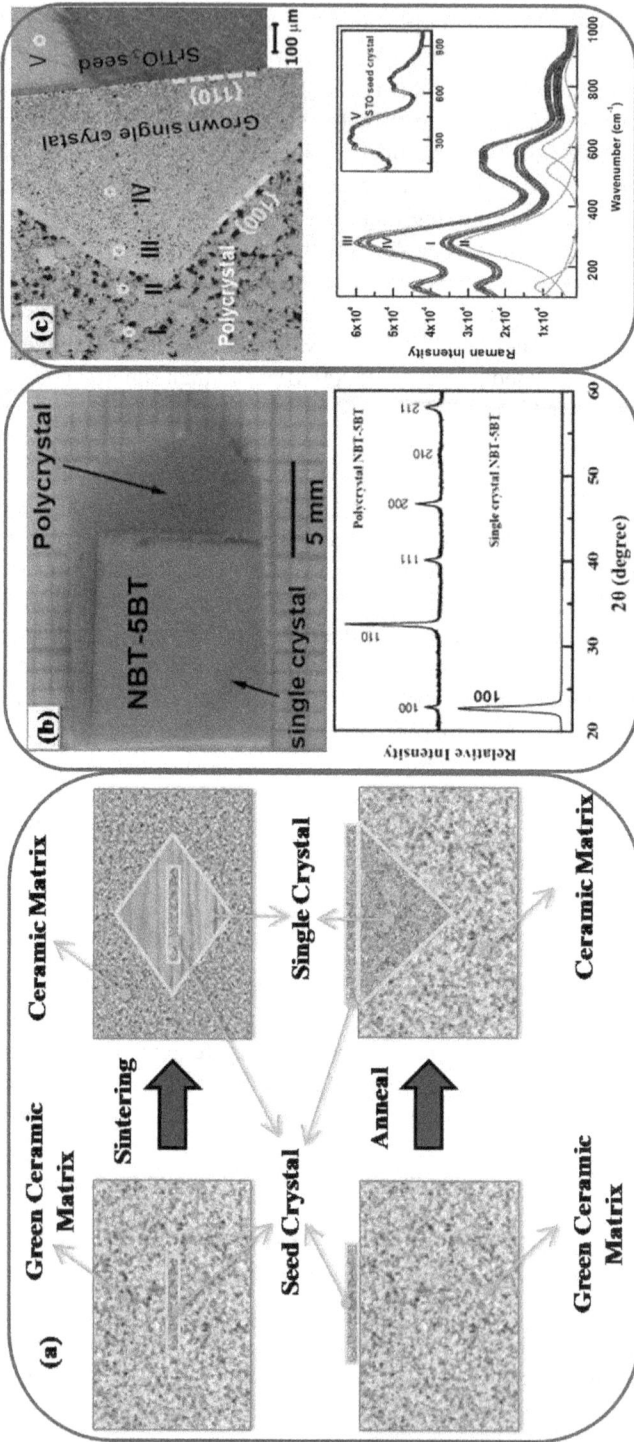

FIGURE 4.7 (a) Single crystal growth mechanism when the seed is embedded inside the green body (top) or placed on top (bottom) of it; (b) single crystal of NBT-BT grown by SSCG method (top) and XRD pattern of the single crystal as compared to the polycrystal (bottom); (c) optical micrograph NBT-BT grown from SrTiO3 seed (top) and Raman spectra of the single crystal taken at three different spots on it (bottom).

Source: [75].

is observed that the piezoelectrically active nanoceramics are embedded in an inactive polymer matrix to form composites for various applications. The orientation of the particles is usually random and requires very high poling fields. In such cases, dielectrophororis is found to be useful. It enhances the piezoelectric constant by controlling the position, alignment, and density of the filler particles as a function of an electric field gradient between the preset electrodes. Khaliq et al. [78] have worked in this direction by using an up-scalable high temperature dielectrophororesis to manufacture 1–3 piezoelectric polymer ceramic composites. Cyclic butylene terephthalate (CBT), a thermoplastic, was used as the matrix while PZT was used as the filler. During dielectrophorosis, the polymer was subjected to a high temperature to facilitate melting, which provided a liquid medium for aligning the PZT particles. The distribution of particles after alignment was frozen upon rapid cooling to room temperature. The reported composites with 10 volume percent of PZT exhibited a piezoelectric voltage sensitivity of 54±4 mV, which was twice that of PZT-based ceramics. In another interesting work, lead-free $BaTiO_3$ nano particles embedded in polydimethylsiloxane (PDMS) elastomer were aligned by dielectrophorosis and employed for blood pressure measurement within a vessel to evaluate the severity of coronary stenosis. To prepare 0–3 oriented BT-PDMS, the composite was heated for 2 hours at 85 °C and subjected to an AC electric field of 3 V/μm, 1 Hz. The applied electric field could orient the sample in both 0–3 and 1–3 directions [79].

4.1.4.2   Injection Molding

Injection molding (Figure 4.8(b)) is a composite manufacturing process that involves heating of a thermoplastic polymer above its melting point and subsequent conversion into molten fluid with reasonably low viscosity. It is then forced through a compact mold defining the design of the expected structure. The polymer is then left to solidify via cooling after which the mold is removed, and the finished part is extracted. This process of plastic molding is not only cost-effective but also can produce large batches of molded structures with high precision. The advantages of plastic injection molding when combined with metal/ceramic powder metallurgy gives rise to a different fabrication process termed as powder injection molding (PIM) for preparing piezoelectric composites. It is useful in fabricating near net shaped metal ceramic structures with complex shapes. Han et al. [80] fabricated a piezoelectric artificial hair cell sensor using the PMN-PZT via PIM process. The sensor manufactured by this process exhibits a high aspect ratio and is utilized for acoustic vector hydrophone along with three mechanoreceptors. Density and dielectric property of the manufactured sensor was 98 percent of the theoretical density and 85 percent of the PMN-PZT powders, respectively. Mechanoreceptors also showed expected piezoelectric property with an 8 percent deviation from the reference. Further, the packaged vector hydrophones were capable of measuring underwater acoustic signals from 500 to 800 Hz with 212 dB of sensitivity. Again, PIM combined with X-ray micromachining was employed to develop high aspect

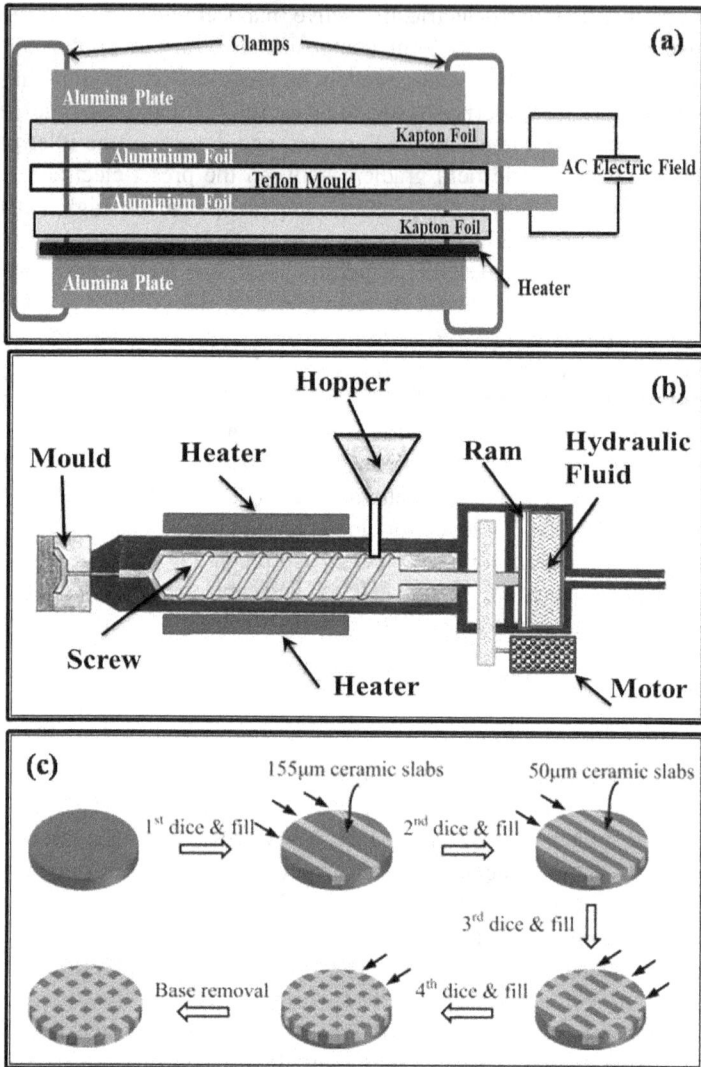

FIGURE 4.8 Basic principle and instrumentation for (a) high temperature dielectrophoresis; (b) injection molding and; (c) dice and fill.

ratio PMN-PZT piezoelectric microarrays. This was an improved technique in the sense that the injection molding conditions were optimized basing on the rheological property of the powder binder mixture to improve incomplete-filling defects. Among the various molding parameters, mold temperature is instrumental in filling up the incomplete defects. The authors of the present work claim the hybrid technique to be instrumental in designing various metals and ceramic applications that require the development of micro-patterns [81]. Similar work was also undertaken by the same author group to fabricate micro PZT arrays in diagnostic 2D array ultrasound transducer [82].

4.1.4.3 Lost Foam Mold

One of the process hazards of injection molding is to keep the green body free from fracturing while demolding. In order to overcome this limitation, the concept of "lost mold" was introduced into ceramic injection molding process. The common approach of all kinds of lost mold techniques includes the use of a sacrificial mold which is burnt out after molding. In the case of the "lost foam" method, a pattern is usually made out of polystyrene foam by a convenient method. To form the pattern, pre-expanded beads of polystyrene are injected into an aluminum mold at low pressure. To cause further expansion, steam is applied, which makes it expand and fill the die. The resulting pattern is only 2.5 percent polystyrene and 97.5 percent air. Further, the foam cluster is coated with the ceramic investment by dipping, brushing, flow coating and spraying. Once the coating dries, the cluster is transferred to a flask with compacted sand. In the process, the ceramic coating captures all the fine details of the mold and acts as a barrier between the mold and sand surface. Not only that, it also controls the permeability, allowing gases from the vaporized foam pattern to escape through the coating into the sand. Finally, the melt is poured into the foam-filled mold, burning the foam out as it pours. The lost mold technique has been used in combination with other techniques to obtain better results. In the past, Guo et al. [83] introduced a new lost rapid prototyping method by combining selective laser sintering (SLS) and gel casting to synthesize piezoelectric ceramics. Basically, the SLS method was used to fabricate sacrificial molds of desired ceramic structure. Aqueous PZT suspension was cast in the mold and solidified in situ via 3D network gel. High strength green body of PZT was obtained, since the polymer mold can be easily removed. The complex PZT structures obtained exhibited optimum electrical properties and were comparable to die pressed PZT sample. Besides, Park et al. [84] fabricated honeycomb shaped, 1–3 piezoelectric micropillar arrays with high aspect ratio using deep X-ray lithography and PIM in six steps: preparation of lost mold, powder binder mixing, injection molding, and demolding, removal of binders and densification of powders. X-ray synchrotron exposure was mainly employed for preparation of polymer-based lost mold insert. Complete filling of the molds was ensured by considering rheological behavior of the piezoelectric powder.

4.1.4.4 Dice and Fill

Dice and fill (Figure 4.8(c)) is a very simple and widespread technique for the fabrication of piezoelectric composites in various applications, especially acoustics. Generally, in this method, composites are prepared by dicing a piezoceramic plate and subsequently filling up of the trenches with interstitial material such as epoxy or any other polymer. Distances between the rods or pixels are so adjusted that lateral modes and other undesirable interference is avoided, as a result of which the distribution of pixels is quite regular and is useful for ultrasound transducer application. In this regard, Mirza et al. [85] synthesized high aspect ratio piezoelectric microrods of $0.9475[Li_{0.02}(K_{0.48}Na_{0.53})_{0.98}](Nb_{0.80}Ta_{0.20})O_3$-$0.0525AgSbO_3$-0.5 wt% MnO_2 (LKNNT-AS-M) embedded in a polymer matrix via the dice-and-fill method. However, previously it was observed that dice-and-fill fabrication of micropillars with large aspect ratio using KNN composites was not easy. This is because microscale pillars tend to collapse during cutting due to the low mechanical strength and fracture toughness of the ceramics. This limitation was overcome by doping with $AgSbO_3$ and MnO_2 which improvise the mechanical strength of KNN. In the beginning of the process, flat surfaces of the discs were arranged parallel to each other by surface grinding and then placed on an automatic dicing machine. At a certain rpm, parallel grooves were made on the disc via a diamond blade (part of the dicing machine). The interstitial spaces were filled with low viscosity epoxy. This process was continued for four sets of discs and, finally, the ceramic base is removed by surface grinding. Further, thin electrodes were deposited on the discs to pole them under a DC electric field of 3.0 kV/ mm at 100 °C. The samples possessed a high thickness factor, $k_t \sim$ 63–67 percent,

lower planer factor $k_p \sim$ 27–34 percent and a superior $\dfrac{k_t}{k_p} \sim$ 2.3. In another work, Xu

et al. [86] employed a modified dice-and-fill method to fabricate magnetoelecctric composites of microscale lead-free piezoceramic pillars incorporated in ferrite matrix without high temperature co-sintering. Previously, ferrites were introduced into the grooves between the pillars by the sol-gel process. but it was observed that the magnetoelectric coefficient of the resulting composites were fairly low due to less ferrite content and incomplete poling of the piezoelectric pillars. However, in the investigation by Xu et al., a high aspect ratio pillar array was constructed on poled piezoceramic plate by a modified dicing method. The discs of poled piezoelectric ceramic were backfilled with wax, which was a very crucial step in modification. Thereafter another set of parallel grooves were cut perpendicular to the former set. Once the scraps and back filled wax were removed, arrays of square shaped ceramic pillars were formed. Subsequently, the $CoFe_2O_4$ nanoparticles were ultrasonically dispersed in ethanol and dropped and deposited into the pillar grooves several times to improve the ferrite content. The samples obtained demonstrated a maximum magnetoelectric response of 302 μV Oe^{-1} with 700 μm high and 150 μm wide pillars.

4.2 SUMMARY

Synthesis methodology plays a very vital role in designing and structuring the piezoelectric materials, either in the form of bulk or nano/ thin films or composites. Therefore, in this chapter, the authors discuss the fabrication techniques suitable for different types of piezoelectric structures and their advancements. At present, the solid state synthesis route is one of the most prominent and commercially used methods for bulk preparation. Similarly, the importance of the sol-gel/hydrothermal method is growing for nanoparticle synthesis due to their uniqueness of developing different microstructures. On the other hand, SSCG is widely used for growing piezoelectric single crystals. Besides, composites, the next-generation piezoelectric materials are preferably fabricated using dice and fill and injection molding. However, synthesis and processing methods can be picked up depending on the requirement of a particular application. This chapter may be helpful in steering research in the direction of adding some cutting-edge technology to the conventional synthesis techniques to obtain piezo-materials with properties tuned for particular applications.

REFERENCES

1. Wang, J., Fan, H., Hu, B. and Jiang, H., 2019. Enhanced energy-storage performance and temperature-stable dielectric properties of $(1-x)(0.94Na_{0.5}Bi_{0.5}TiO_3-0.06BaTiO_3)-xNa_{0.73}Bi_{0.09}NbO_3$ ceramics. *Journal of Materials Science: Materials in Electronics*, *30*(3), pp. 2479–2488.

2. Carter, R.E., 1961. Kinetic model for solid-state reactions. *The Journal of Chemical Physics*, *34*(6), pp. 2010–2015.

3. Brzozowski, E. and Castro, M.S., 2000. Synthesis of barium titanate improved by modifications in the kinetics of the solid state reaction. *Journal of the European Ceramic Society*, *20*(14–15), pp. 2347–2351.

4. Awan, I.T., Pinto, A.H., Nogueira, I.C., Bezzon, V.D.N., Leite, E.R., Balogh, D.T., Mastelaro, V.R., Ferreira, S.O. and Marega Jr, E., 2020. Insights on the mechanism of solid state reaction between TiO_2 and $BaCO_3$ to produce $BaTiO_3$ powders: The role of calcination, milling, and mixing solvent. *Ceramics International*, *46*(3), pp. 2987–3001.

5. Kainz, T., Naderer, M., Schütz, D., Fruhwirth, O., Mautner, F.A. and Reichmann, K., 2014. Solid state synthesis and sintering of solid solutions of BNT–xBKT. *Journal of the European Ceramic Society*, *34*(15), pp. 3685–3697.

6. Swartz, S.L. and Shrout, T.R., 1982. Fabrication of perovskite lead magnesium niobate. *Materials Research Bulletin*, *17*(10), pp. 1245–1250.

7. Liou, Y.C. and Chen, J.H., 2004. PMN ceramics produced by a simplified columbite route. *Ceramics International*, *30*(1), pp. 17–22.

8. Bruno, J.C., Cavalheiro, A.A., Zaghete, M.A., Cilense, M. and Varela, J.A., 2006. Characterization of the columbite precursor and (1-x) PMN-xPT powders prepared by Ti-modified columbite route. *Ferroelectrics*, *339*(1), pp. 227–234.

9. Amer, A.M., Ibrahim, S.A., Ramadan, R.M. and Ahmed, M.S., 2005. Reactive calcination derived PZT ceramics. *Journal of Electroceramics*, *14*(3), pp. 273–281.

10. Rout, D., Subramanian, V., Hariharan, K., Sivasubramanian, V. and Murthy, V.R.K., 2004. Dielectric Study of the Phase Transitions in $Pb_{1-x}Ba_x(Yb_{1/2}Ta_{1/2})O_3$ Ceramics. *Ferroelectrics*, *300*(1), pp. 67–78.

11. Rout, D., Subramanian, V., Hariharan, K., Murthy, V.R.K. and Sivasubramanian, V., 2005. Raman spectroscopic study of $(Pb_{1-x}Ba_x)(Yb_{1/2}Ta_{1/2})O_3$ ceramics. *Journal of Applied Physics*, *98*(10), p. 103503.

12. Rout, D., Subramanian, V., Hariharan, K. and Sivasubramanian, V., 2005. Microtructure and dielectric properties of $(1-x)Pb(Yb_{1/2}Ta_{1/2})O_3-xPb(Fe_{1/2}Ta_{1/2})O_3$,

$0 \leq x \leq 0.2$ solid solution ceramics. *Materials Science and Engineering: B, 123*(2), pp. 107–114.

13. Gao, X., Wu, J., Yu, Y., Chu, Z., Shi, H. and Dong, S., 2018. Giant piezoelectric coefficients in relaxor piezoelectric ceramic PNN-PZT for vibration energy harvesting. *Advanced Functional Materials, 28*(30), p. 1706895.

14. Danks, A.E., Hall, S.R. and Schnepp, Z.J.M.H., 2016. The evolution of 'sol–gel' chemistry as a technique for materials synthesis. *Materials Horizons, 3*(2), pp. 91–112.

15. Luginbuhl, P., Racine, G.A., Lerch, P., Romanowicz, B., Brooks, K.G., De Rooij, N.F., Renaud, P. and Setter, N., 1996. Piezoelectric cantilever beams actuated by PZT sol-gel thin film. *Sensors and Actuators A: Physical, 54*(1–3), pp. 530–535.

16. Kholkin, A.L., Yarmarkin, V.K., Wu, A., Avdeev, M., Vilarinho, P.M. and Baptista, J.L., 2001. PZT-based piezoelectric composites via a modified sol–gel route. *Journal of the European Ceramic Society, 21*(10–11), pp. 1535–1538.

17. Cernea, M., Monnereau, O., Llewellyn, P., Tortet, L. and Galassi, C., 2006. Sol-gel synthesis and characterization of Ce doped-$BaTiO_3$. *Journal of the European Ceramic Society, 26*(15), pp. 3241–3246.

18. Cernea, M., Andronescu, E., Radu, R., Fochi, F. and Galassi, C., 2010. Sol–gel synthesis and characterization of $BaTiO_3$-doped $(Bi_{0.5}Na_{0.5})TiO_3$ piezoelectric ceramics. *Journal of Alloys and Compounds, 490*(1–2), pp. 690–694.

19. Dashtizad, S., Alizadeh, P. and Yourdkhani, A., 2021. Improving piezoelectric properties of PVDF fibers by compositing with $BaTiO_3$-Ag particles prepared by sol-gel method and photochemical reaction. *Journal of Alloys and Compounds, 883*, p. 160810.

20. Simon-Seveyrat, L., Hajjaji, A., Emziane, Y., Guiffard, B. and Guyomar, D., 2007. Re-investigation of synthesis of $BaTiO_3$ by conventional solid-state reaction and oxalate coprecipitation route for piezoelectric applications. *Ceramics International, 33*(1), pp. 35–40.

21. Batra, K., Sinha, N., Goel, S., Yadav, H., Joseph, A.J. and Kumar, B., 2018. Enhanced dielectric, ferroelectric and piezoelectric performance of Nd-ZnO nanorods and their application in flexible piezoelectric nanogenerator. *Journal of Alloys and Compounds, 767*, pp. 1003–1011.

22. Sánchez-Hernández, A.K., Lozano-Rosas, R., Gervacio-Arciniega, J.J., Wang, J. and Robles-Águila, M.J., 2022. Piezoelectric and mechanical properties of hydroxyapatite/titanium oxide composites. *Ceramics International, 48*, pp. 23096–23103.

23. Yang, Z., Chang, Y. and Li, H., 2005. Piezoelectric and dielectric properties of PZT–PZN–PMS ceramics prepared by molten salt synthesis method. *Materials Research Bulletin, 40*(12), pp. 2110–2119.

24. Ge, H., Hou, Y., Xia, C., Zhu, M., Wang, H. and Yan, H., 2011. Preparation and piezoelectricity of $NaNbO_3$ high-density ceramics by molten salt synthesis. *Journal of the American Ceramic Society, 94*(12), pp. 4329–4334.

25. Gupta, S.K. and Mao, Y., 2021. Recent developments on molten salt synthesis of inorganic nanomaterials: a review. *The Journal of Physical Chemistry C, 125*(12), pp. 6508–6533.

26. Xiao, X., Wang, H., Urbankowski, P. and Gogotsi, Y., 2018. Topochemical synthesis of 2D materials. *Chemical Society Reviews, 47*(23), pp. 8744–8765.

27. Roy, B., Ahrenkiel, S.P. and Fuierer, P.A., 2008. Controlling the size and morphology of TiO2 powder by molten and solid salt synthesis. *Journal of the American Ceramic Society, 91*(8), pp. 2455–2463.

28. Zhou, X., Wu, Z., Jiang, C., Luo, H., Yan, Z. and Zhang, D., 2018. Molten salt synthesis and characterization of lead-free $(1-x)Na_{0.5}Bi_{0.5}TiO_3$-$x$SrTiO$_3$ ($x = 0$, 0.10, 0.26) whiskers. *Ceramics International*, *44*(8), pp. 9174–9180.

29. Huang, G., Lu, C.H. and Yang, H.H., 2019. Magnetic nanomaterials for magnetic bioanalysis. In *Novel Nanomaterials for Biomedical, Environmental and Energy Applications* (pp. 89–109). Elsevier.

30. Maeda, T., Takiguchi, N., Ishikawa, M., Hemsel, T. and Morita, T., 2010. (K, Na) NbO$_3$ lead-free piezoelectric ceramics synthesized from hydrothermal powders. *Materials Letters*, *64*(2), pp. 125–128.

31. Zhou, X., Jiang, C., Chen, C., Luo, H., Zhou, K. and Zhang, D., 2016. Morphology control and piezoelectric response of Na 0.5 Bi 0.5 TiO 3 synthesized via a hydrothermal method. *CrystEngComm*, *18*(8), pp. 1302–1310.

32. Tan, L., Wang, X., Zhu, W., Li, A. and Wang, Y., 2021. Excellent piezoelectric performance of KNNS-based lead-free piezoelectric ceramics through powder pretreatment by hydrothermal method. *Journal of Alloys and Compounds*, *874*, p. 159770.

33. Jung, D.S., Ko, Y.N., Kang, Y.C. and Park, S.B., 2014. Recent progress in electrode materials produced by spray pyrolysis for next-generation lithium ion batteries. *Advanced Powder Technology*, *25*(1), pp. 18–31.

34. Haugen, A.B., Madaro, F., Bjørkeng, L.P., Grande, T. and Einarsrud, M.A., 2015. Sintering of sub-micron K$_{0.5}$Na$_{0.5}$NbO$_3$ powders fabricated by spray pyrolysis. *Journal of the European Ceramic Society*, *35*(5), pp. 1449–1457.

35. Ayana, A., Hou, F., Seidel, J., Rajendra, B.V. and Sharma, P., 2022. Microstructural and piezoelectric properties of ZnO films. *Materials Science in Semiconductor Processing*, *146*, p. 106680.

36. Ramos-Serrano, J.R., Alcántara-Iniesta, S., Acosta-Osorno, M. and Calixto, M.E., 2022. Growth of highly c-axis oriented ZnO thin films by spray pyrolysis for piezoelectric applications. *Materials Science in Semiconductor Processing*, *144*, p. 106585.

37. Kim, B.H., Han, S.J., Kim, J.H., Lee, J.H., Ahn, B.K. and Xu, Q., 2007. Electrical properties of $(1-x)(Bi_{0.5}Na_{0.5})TiO_3$-$x$BaTiO$_3$ synthesized by emulsion method. *Ceramics International*, *33*(3), pp. 447–452.

38. Rossnagel, S.M., 2003. Thin film deposition with physical vapor deposition and related technologies. *Journal of Vacuum Science & Technology A: Vacuum, Surfaces, and Films*, *21*(5), pp. S74–S87.

39. Krupanidhi, S.B. and Sayer, M., 1984. Position and pressure effects in rf magnetron reactive sputter deposition of piezoelectric zinc oxide. *Journal of Applied Physics*, *56*(11), pp. 3308–3318.

40. Wacogne, B., Roe, M.P., Pattinson, T.J. and Pannell, C.N., 1995. Effective piezoelectric activity of zinc oxide films grown by radio-frequency planar magnetron sputtering. *Applied Physics Letters*, *67*(12), pp. 1674–1676.

41. Ababneh, A., Schmid, U., Hernando, J., Sánchez-Rojas, J.L. and Seidel, H., 2010. The influence of sputter deposition parameters on piezoelectric and mechanical properties of AlN thin films. *Materials Science and Engineering: B*, *172*(3), pp. 253–258.

42. Cao, C.B., Chan, H.L.W. and Choy, C.L., 2003. Piezoelectric coefficient of InN thin films prepared by magnetron sputtering. *Thin Solid Films*, *441*(1–2), pp. 287–291.

43. Yanagitani, T. and Kiuchi, M., 2007. Control of in-plane and out-of-plane texture in shear mode piezoelectric ZnO films by ion-beam irradiation. *Journal of Applied Physics*, *102*(4), p. 044115.

44. Agarwal, D.C., Chauhan, R.S., Kumar, A., Kabiraj, D., Singh, F., Khan, S.A., Avasthi, D.K., Pivin, J.C., Kumar, M., Ghatak, J. and Satyam, P.V., 2006. Synthesis and characterization of ZnO thin film grown by electron beam evaporation. *Journal of Applied Physics*, *99*(12), p. 123105.

45. Periasamy, C. and Chakrabarti, P., 2011. Time-dependent degradation of Pt/ZnO nanoneedle rectifying contact based piezoelectric nanogenerator. *Journal of Applied Physics*, *109*(5), p. 054306.

46. Sun, L., Yuan, G., Gao, L., Yang, J., Chhowalla, M., Gharahcheshmeh, M.H., Gleason, K.K., Choi, Y.S., Hong, B.H. and Liu, Z., 2021. Chemical vapour deposition. *Nature Reviews Methods Primers*, *1*(1), pp. 1–20.

47. Kim, N.I., Chen, J., Wang, W., Moradnia, M., Pouladi, S., Kwon, M.K., Kim, J.Y., Li, X. and Ryou, J.H., 2021. Highly-sensitive skin-attachable eye-movement sensor using flexible nonhazardous piezoelectric thin film. *Advanced Functional Materials*, *31*(8), p. 2008242.

48. Cuniot-Ponsard, M., Saraswati, I., Ko, S.M., Halbwax, M., Cho, Y.H. and Dogheche, E., 2014. Electro-optic and converse-piezoelectric properties of epitaxial GaN grown on silicon by metal-organic chemical vapor deposition. *Applied Physics Letters*, *104*(10), p. 101908.

49. Yang, S.Y., Zavaliche, F., Mohaddes-Ardabili, L., Vaithyanathan, V., Schlom, D.G., Lee, Y.J., Chu, Y.H., Cruz, M.P., Zhan, Q., Zhao, T. and Ramesh, A.R., 2005. Metalorganic chemical vapor deposition of lead-free ferroelectric $BiFeO_3$ films for memory applications. *Applied Physics Letters*, *87*(10), p. 102903.

50. Sánchez, G., Abdallah, B., Tristant, P., Dublanche-Tixier, C., Djouadi, M.A., Besland, M.P., Jouan, P.Y. and Bologna Alles, A., 2009. Microstructure and mechanical properties of AlN films obtained by plasma enhanced chemical vapor deposition. *Journal of Materials Science*, *44*(22), pp. 6125–6134.

51. Goff, E.D. and Braymen, S.D., 1994. A novel approach to the deposition of piezoelectric thin films. *Materials Letters*, *21*(3–4), pp. 347–349.

52. Volintiru, I., Creatore, M., Kniknie, B.J., Spee, C.I.M.A. and Van De Sanden, M.C.M., 2007. Evolution of the electrical and structural properties during the growth of Al doped ZnO films by remote plasma-enhanced metalorganic chemical vapor deposition. *Journal of Applied Physics*, *102*(4), p. 043709.

53. Dutta, M., Mridha, S. and Basak, D., 2008. Effect of sol concentration on the properties of ZnO thin films prepared by sol–gel technique. *Applied Surface Science*, *254*(9), pp. 2743–2747.

54. Cernea, M., Trupina, L., Dragoi, C., Vasile, B.S. and Trusca, R., 2012. Structural and piezoelectric characteristics of BNT–BT0.05 thin films processed by sol–gel technique. *Journal of Alloys and Compounds*, *515*, pp. 166–170.

55. Li, W., Zeng, H., Hao, J. and Zhai, J., 2013. Enhanced dielectric and piezoelectric properties of Mn doped $(Bi_{0.5}Na_{0.5})TiO_3$–$(Bi_{0.5}K_{0.5})TiO_3$–$SrTiO_3$ thin films. *Journal of Alloys and Compounds*, *580*, pp. 157–161.

56. Yoo, S.E., Hayashi, M., Ishizawa, N. and Yoshimura, M., 1990. Preparation of Strontium Titanate Thin Film on Titanium Metal Substrate by Hydrothermal-Electrochemical Method. *Journal of the American Ceramic Society*, *73*(8), pp. 2561–2563.

57. Lu, F.H., Wu, C.T. and Hung, C.Y., 2002. Barium titanate films synthesized by an anodic oxidation-based electrochemical method. *Surface and Coatings Technology*, *153*(2–3), pp. 276–283.
58. Tamvakos, D., Lepadatu, S., Antohe, V.A., Tamvakos, A., Weaver, P.M., Piraux, L., Cain, M.G. and Pullini, D., 2015. Piezoelectric properties of template-free electrochemically grown ZnO nanorod arrays. *Applied Surface Science*, *356*, pp. 1214–1220.
59. Seo, K.W. and Kong, H.G., 2000. Hydrothermal preparation of $BaTiO_3$ thin films. *Korean Journal of Chemical Engineering*, *17*(4), pp. 428–432.
60. Padture, N.P. and Wei, X., 2003. Hydrothermal Synthesis of Thin Films of Barium Titanate Ceramic Nano-Tubes at 200° C. *Journal of the American Ceramic Society*, *86*(12), pp. 2215–2217.
61. Shimomura, K., Tsurumi, T., Ohba, Y.O.Y. and Daimon, M.D.M., 1991. Preparation of lead zirconate titanate thin film by hydrothermal method. *Japanese Journal of Applied Physics*, *30*(9S), p. 2174.
62. Scheel, H.J., 2007. Introduction to liquid phase epitaxy. *Liquid Phase Epitaxy of Electronic, Optical and Optoelectronic Materials*, *19*(8).
63. Zhang, Q., Sánchez-Fuentes, D., Gómez, A., Desgarceaux, R., Charlot, B., Gàzquez, J., Carretero-Genevrier, A. and Gich, M., 2019. Tailoring the crystal growth of quartz on silicon for patterning epitaxial piezoelectric films. *Nanoscale Advances*, *1*(9), pp. 3741–3752.
64. Xu, C., Li, Q., Yan, Q., Luo, N., Zhang, Y. and Chu, X., 2016. Domain Structure Evolutions During the Poling Process for [011]-Oriented PMN–xPT Crystals Across the MPB Region. *Journal of the American Ceramic Society*, *99*(6), pp. 2096–2102.
65. Yu, F.P., Yuan, D.R., Yin, X., Zhang, S.J., Pan, L.H., Guo, S.Y., Duan, X.L. and Zhao, X., 2009. Czochralski growth and characterization of the piezoelectric single crystal La3Ga5. 5Nb0. 5O14. *Solid State Communications*, *149*(31–32), pp. 1278–1281.
66. Mancini, G.F., Ghigna, P., Mozzati, M.C., Galinetto, P., Makarova, M., Syrnikov, P., Jastrabik, L. and Trepakov, V.A., 2014. Structural investigation of manganese doped SrTiO3 single crystal and ceramic. *Ferroelectrics*, *463*(1), pp. 31–39.
67. Souleiman, M., Bhalerao, G.M., Guillet, T., Haidoux, A., Cambon, M., Levelut, C., Haines, J. and Cambon, O., 2014. Hydrothermal growth of large piezoelectric single crystals of GaAsO4. *Journal of Crystal Growth*, *397*, pp. 29–38.
68. Cambon, O., Bhalerao, G.M., Bourgogne, D., Haines, J., Hermet, P., Keen, D.A. and Tucker, M.G., 2011. Vibrational origin of the thermal stability in the high-performance piezoelectric material GaAsO4. *Journal of the American Chemical Society*, 133(20), pp. 8048–8056.
69. Jethva, H.O., 2017. Gel growth: a brief review. *Mechanics, Materials Science & Engineering MMSE Journal. Open Access, 9*, ff10.2412/mmse.64.79.613ff. ffhal-01504659f.
70. Elwell, D., 1989. Fundamentals of flux growth. In, *Crystal Growth in Science and Technology*, pp. 133–142. Springer, Boston.
71. Yi, X., Chen, H., Cao, W., Zhao, M., Yang, D., Ma, G., Yang, C. and Han, J., 2005. Flux growth and characterization of lead-free piezoelectric single crystal $[Bi_{0.5}(Na_{1-x}K_x)_{0.5}]TiO_3$. *Journal of Crystal Growth*, *281*(2–4), pp. 364–369.
72. Singh, B.K., Kumar, K., Sinha, N. and Kumar, B., 2009. Flux growth and low temperature dielectric relaxation in piezoelectric $Pb[(Zn_{1/3}Nb_{2/3})_{0.91}Ti_{0.09}]O_3$ single crystals. *Crystal Research and Technology: Journal of Experimental and Industrial Crystallography*, *44*(9), pp. 915–924.

73. Dhanaraj, G., Byrappa, K., Prasad, V.V. and Dudley, M., 2010. Crystal growth techniques and characterization: an overview. *Springer Handbook of Crystal Growth*, pp. 3–16.

74. Milisavljevic, I., and Wu, Y., 2020. Current status of solid-state single crystal growth. *BMC Materials*, *2*(1), pp. 1–26.

75. Moon, K.S., and Kang, S.J.L., 2008. Coarsening behavior of round-edged cubic grains in the $Na_{1/2}Bi_{1/2}TiO_3$–$BaTiO_3$ system. *Journal of the American Ceramic Society*, *91*(10), pp. 3191–3196.

76. Moon, K.S., Rout, D., Lee, H.Y. and Kang, S.J.L., 2011. Solid state growth of $Na_{1/2}Bi_{1/2}TiO_3$–$BaTiO_3$ single crystals and their enhanced piezoelectric properties. *Journal of Crystal Growth*, *317*(1), pp. 28–31.

77. Moon, K.S., Rout, D., Lee, H.Y. and Kang, S.J.L., 2011. Effect of TiO_2 addition on grain shape and grain coarsening behavior in $95Na_{1/2}Bi_{1/2}TiO_3$–$5BaTiO_3$. *Journal of the European Ceramic Society*, *31*(10), pp. 1915–1920.

78. Khaliq, J., Hoeks, T. and Groen, P., 2019. Fabrication of piezoelectric composites using high-temperature dielectrophoresis. *Journal of Manufacturing and Materials Processing*, *3*(3), p. 77.

79. d'Ambrogio, G., Zahhaf, O., Bordet, M., Le, M.Q., Della Schiava, N., Liang, R., Cottinet, P.J. and Capsal, J.F., 2021. Structuring $BaTiO_3$/PDMS nanocomposite via dielectrophoresis for fractional flow reserve measurement. *Advanced Engineering Materials*, *23*(10), p. 2100341.

80. Han, J.S., Oh, K.H., Moon, W.K., Kim, K., Joh, C., Seo, H.S., Bollina, R. and Park, S.J., 2015. Bio-inspired piezoelectric artificial hair cell sensor fabricated by powder injection molding. *Smart Materials and Structures*, *24*(12), p. 125025.

81. Han, J.S., Gal, C.W., Kim, J.H. and Park, S.J., 2016. Fabrication of high-aspect-ratio micro piezoelectric array by powder injection molding. *Ceramics International*, *42*(8), pp. 9475–9481.

82. Han, J.S., Gal, C.W., Park, J.M., Kim, J.H., Lee, S.H., Yeo, B.W., Lee, B.W., Park, S.S. and Park, S.J., 2018. Powder injection molding process for fabrication of piezoelectric 2D array ultrasound transducer. *Smart Materials and Structures*, *27*(7), p. 075058.

83. Guo, D., Li, L.T., Cai, K., Gui, Z.L. and Nan, C.W., 2004. Rapid prototyping of piezoelectric ceramics via selective laser sintering and gelcasting. *Journal of the American Ceramic Society*, *87*(1), pp. 17–22.

84. Park, J.M., Han, J.S., Oh, J.W., Park, S.C., Kim, Y.D., Jang, J.M., Lee, W.S. and Park, S.J., 2020. Design, fabrication of honeycomb-shaped 1–3 connectivity piezoelectric micropillar arrays for 2D ultrasound transducer application. *Ceramics International*, *46*(8), pp. 12023–12030.

85. Mirza, M.S., Liu, Q., Yasin, T., Qi, X., Li, J.F. and Ikram, M., 2016. Dice-and-fill processing and characterization of microscale and high-aspect-ratio $(K,Na)NbO_3$-based 1–3 lead-free piezoelectric composites. *Ceramics International*, *42*(9), pp. 10745–10750.

86. Xu, Y., Li, J.F., Ma, J. and Nan, C.W., 2012. Microscale 1–3-type lead-free piezoelectric/ferrite composites fabricated by a modified dice-and-fill method. *Journal of Physics D: Applied Physics*, *45*(31), p. 315306.

5 Characterization and Properties of Piezoelectric Materials

5.1 CHARACTERIZATION TECHNIQUES

5.1.1 X-ray Diffraction

In the field of material science, X-ray diffraction has grown in strength for decades. Diffraction has been used as a characterization method for almost a century, and the discipline is always evolving with new technologies constantly permitting additional characterization dimensions. In all areas of materials research, including the study of piezoelectric materials, diffraction is used most frequently to characterize crystal structure, phase history, particle size, crystallographic texture, and lattice strains. Recent developments in diffraction technology have made it possible to solve new issues in this research field. For instance, the development of microdiffraction may be able to identify structural flaws like domain walls and cracks. Specifically with the advancement of time-resolved approaches, ferroelectrics can now be characterized in real-time when cyclic electric fields are applied. Development in the characterization techniques for study of piezoelectric materials and conventional applications of X-ray diffraction are getting explored. The opportunity to comprehend the piezoelectric action in polycrystalline bulk ceramics is provided by these new characterization techniques. Numerous present issues in this area will be resolved by adopting a combinatorial characterization approach and the connection of diffraction observations to theoretical models [1].

As mentioned, XRD is primarily used to characterize piezoelectric materials for structure determination by evaluating multiple diffraction peaks as it contains more crystallographic information. Diffraction of the X-Rays through the crystals follows Bragg's law of diffraction (Figure 5.1):

$$2d \sin \theta = n\lambda$$

where d is the inter-planer spacing of the crystal, θ is the angle of diffraction, λ is the incident X-Ray beam wavelength and n is the order of diffraction. The intensity of the diffracted X-ray beam depends on how the atoms are arranged within the unit cell and also on the atomic number of the respective ions which in turn describe the

DOI: 10.1201/9781003317289-5

symmetry of the structures. A standard XRD pattern is shown in Figure 5.1. Further, Rietveld technique is applied by many researches for refining the structures. Again, phase evolution in piezo-materials is quite interesting before and after processing; and before and after applying electric fields or stress. Particularly during electrical or mechanical loading, morphotropic phase boundary (MPB) compositions can show field-induced phase alterations. For the first time, Noheda et al. [2, 3] evidenced a monoclinic phase in $Pb(Zr_xTi_{1-x})O_3$ near the MPB and later this phase was also found existing at ambient temperature when the sample was polled under high electric field. To comprehend the connection between the crystal structure and the piezoelectric properties, Kim et al. [4] studied $(1-x)(Bi_{0.5}Na_{0.5})TiO_3$-$xSrTiO_3$ (BNT-xST) piezoelectric ceramic system. Based on the structural refinement analysis by Rietveld method (Figure 5.2(a)-(b)), they have shown that the observed piezoelectric constant d_{33} is connected with the weighted off-center values (d_w) and the phase weighted cation-anion average distances; ceramics with higher d_w values have higher d_{33}. As the $SrTiO_3$ concentration increases, the crystal symmetry gradually shifts from rhombohedral ($x = 0.00$) to rhombohedral-tetragonal ($x = 0.10$–0.30), then to tetragonal-cubic ($x = 0.40$–0.60), and finally to cubic ($x = 0.80$–1.00) phases (Table 5.1). Earlier in 2010, Rout et al. [5] studied the temperature–and composition-driven phase transitions in the same NBT-xST ($0 \le x \le 40$) by X-ray diffraction in correlation with dielectric, and Raman spectroscopy over a wide temperature range (20–550° C). Similarly, the rhombohedral-tetragonal MPB phase coexistence in $(1-x)(Bi_{0.5}Na_{0.5})TiO_3$-$xBaTiO_3$ (BNT-xBT) was verified around $x = 5.5$ by the same group using temperature dependent XRD [6]. Utilizing XRD and Rietveld refinement, Habib et al. [7] have observed two composition driven MPB and lattice distortion (c_T/a_T and $90°-\alpha_R$) in a lead-free system $0.7Bi_{1.03(1-x)}La_xFeO_3$-$0.3BaTiO_3$ and correlated the structural evolution with enhanced piezoelectric properties. To learn more about the relative contributions of intrinsic and extrinsic influences to the origins of high piezo-properties in $(K_{0.44}Na_{0.52}Li_{0.04})(Nb_{0.86}Ta_{0.10}Sb_{0.04})$ O_3 polycrystalline ceramics, Ochoa et al. [8] calculated the non-180° domain reorientation and lattice strain from high-energy X-ray diffraction data. They have

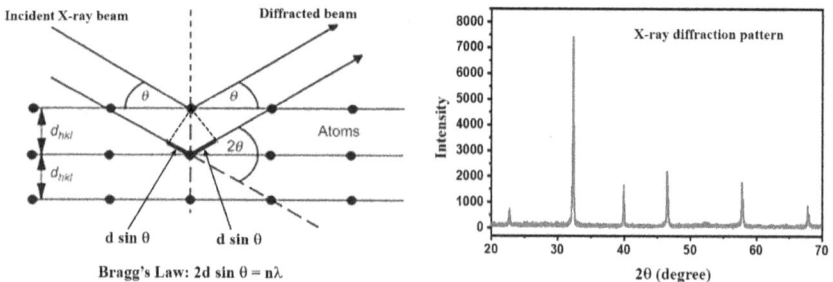

FIGURE 5.1 Schematic illustration showing diffraction of X-rays from lattice planes following Bragg's law and a standard X-Ray spectrum.

TABLE 5.1
Fitting parameters after Rietveld refinement

STO content (x)	Phases used in refinement	wR$_B$ X-ray (%)	R$_{WP}$ X-ray (%)	χ^2 (%)
0.00	Rhombohedral	5.51	6.40	3.36
0.10	Rhombohedral + Tetragonal	6.60	6.93	3.40
0.15	Rhombohedral + Tetragonal	6.12	6.80	3.83
0.20	Rhombohedral + Tetragonal	5.35	6.04	2.42
0.22	Rhombohedral + Tetragonal	5.50	6.25	3.63
0.25	Rhombohedral + Tetragonal	5.56	6.72	2.72
0.30	Rhombohedral + Tetragonal	7.34	7.36	3.12
0.40	Tetragonal + Cubic	6.98	7.14	2.76
0.60	Tetragonal + Cubic	5.62	6.52	2.31
0.80	Cubic	6.71	6.92	2.25
1.00	Cubic	5.73	6.58	2.76

Source: [4].

recorded the 2D diffraction pattern in presence of applied electric field using an experimental set up (schematic diagram) given in Figure 5.2(c). Figure 5.2(d)-(f) shows the domain switching (intensity variation of 200 Bragg peak with respect to applied electric field) and the analysis as function of electric field and angle to the applied field. Moon et al. [9] confirmed the growth of single crystal of NBT-5BT piezoelectric with the (100) plane on a (110) $SrTiO_3$ seed crystal using micro-area X-ray diffraction. Figure 5.3(a) clearly distinguishes the poly-crystalline and single crystal part of the crystal growth. Rout et al. [10] have successfully grown a heteroepitaxial $BiFeO_3$ film on a $SrTiO_3$ substrate (100-oriented single-crystal) by hydrothermal epitaxy at a very low temperature 200 °C and the epitaxial growth (cube-on-cube) was confirmed from the results of Pole-figure and F-scan.

Moreover, the texture measurement in polycrystalline ceramics has long relied on diffraction. The goal of crystallographic texture in piezoelectrics is to provide anisotropic properties, which typically results in an improved piezo-response in a preferred direction as compared to a randomly oriented ceramic. Usually, the texture is induced either to the ferroelectric domain structure (domain texture) through poling or to the grain orientations (grain texture) through uniaxial pressing, tape casting, hot forging, and so forth. Rout et al. [11] fabricated textured $CaBi_4Ti_4O_{15}$ (CBT) ceramics by hot-forging with a degree of orientation (F) along the crystal structure's c-axis is 89.4 percent calculated using XRD pattern (Figure 5.3(b)). They observed significant anisotropy having the ratio of CBT ($\perp$) to CBT (II) permittivity as 3 at transition temperature. Similarly, Hussain et al. [12] have fabricated $Bi_{0.5}(Na_{0.75}K_{0.25})_{0.5}TiO_3$ (BNKT) textured ceramics by reactive templated grain growth (TGG) method and studied the effect of grain orientation on structure, dielectric, and electrical properties.

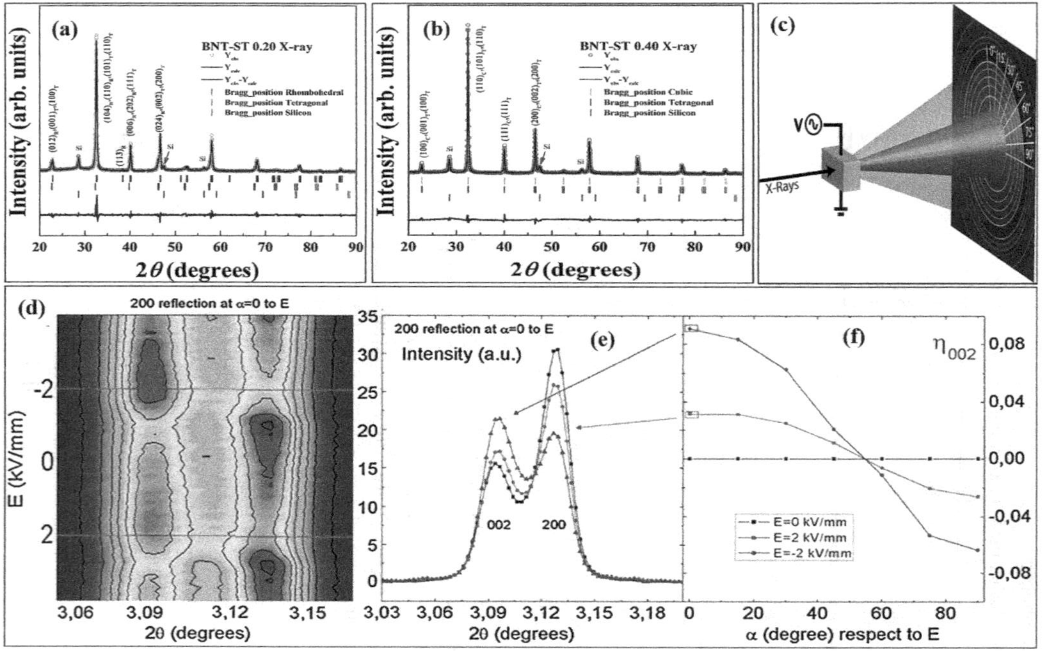

FIGURE 5.2 (a) and (b) Reitveld refinement results to confirm structural coexistence in BNT-0.2ST and BNT-0.4ST ceramics respectively; (c) schematic diagram showing experimental set up of 2D diffraction in presence electric field; (d) 200 reflection contour on application of electric field where domain switching is represented by intensity differences; (e) domain switching analysis for comparing the peak intensities for different applied electric fields; (f) domain switching (η002) as a function of angle to the applied electric field.

Source: [4, 8].

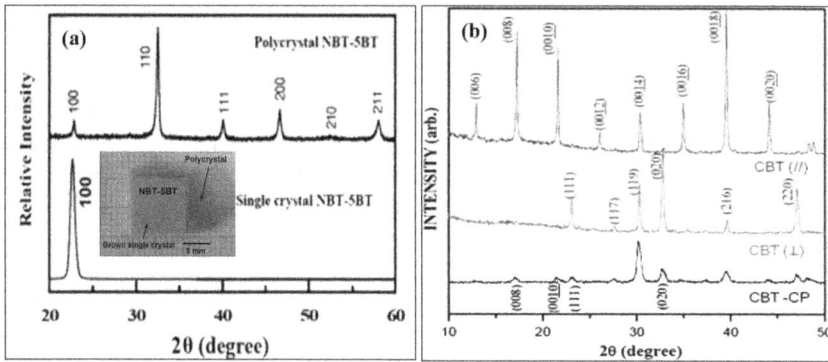

FIGURE 5.3 (a) Typical XRD pattern of polycrystalline and single crystal of NBT-5BT; (b) room temperature X-Ray diffraction patterns of CBT-CP, CBT (⊥) and CBT(∥).

Source: [9, 11].

5.1.2 DENSITY

For more than the last five decades researchers worldwide have established that the eletromechanical properties of piezoceramics inherently rely on their chemical bonding, composition, phase, crystal structure, microstructure, density, flaw size, indentation area and so forth. The formation of an undesirable phase (secondary/impurity phase), non-stoichiometry, a defective microstructure, insufficient densification, and other defects are all very likely to occur during the processing of ceramics if proper caution is not taken in procedures like powder synthesis, shaping, and sintering. Furthermore, it is important to realize that even if ceramic powder using high purity chemicals is synthesized with desired phase, still attention needs to be paid to produce dense ceramics. This is essential in order to ensure required performance at different operational conditions like high temperature, pressure and electric fields. The basic understanding of bulk density depicts the ratio of the entire mass of a body to its bulk volume (i.e., the total area inside the body's macroscopic "envelope" surface). Usually, the bulk density is measured using Archimedes' principle or by image processing approach. However, Archimedes' principle is mostly used for density measurement of piezoceramics. To calculate the density, the formula used is:

$$\text{Density}(\rho) = \frac{W1}{W3 - W2} \times \rho w$$

Where $W1$ = the weight of dry specimen, $W2$ = the weight of the specimen in water, $W3$ is the water saturated weight (i.e., the weight of the sample measured in air after taking out from water) and ρ_w is the density of water.

Generally, in ceramic engineering, the most conventional method to produce a dense ceramic is to form a green body (pressing the synthesized ceramic powders and various organic or inorganic binders), followed by high temperature sintering. A powder compact (green body/pellet) is subjected to a heat treatment process using a furnace called sintering in order to add strength and integrity. Practically, the bulk volume that includes the volume of solid particles, additives/binders, liquid phase, and empty pore space reduces significantly during sintering. The sintering temperature is always kept below the melting point of the major components of the powder compact. Depending on materials, the densification can be maximized by optimizing the sintering temperature and sintering time, pressure (in case of pressure assisted sintering), additives (in case of liquid phase sintering) and so forth and subsequently improves the poling treatment for property enhancement. Also, to improve densification and piezoelectric properties, suitable sintering techniques like normal sintering in air or other mediums, hot-pressing, spark plasma sintering and so forth were employed. Many researchers have effectively correlated the bulk density with piezoelectric property enhancement. For example: $K_{0.5}Na_{0.5}NbO_3$ (KNN) ceramic has a long-term issue in densification because of the intrinsic volatility of alkaline elements during conventional sintering [13]. However, Jaeger and Egerton [14] succeeded in producing relatively high density ceramics using pressure-assisted hot-pressing method with high piezoelectric constant (d_{33}) and Curie temperature (T_C). Making use of the advantages of rapid heating rate and short soaking time, Wang et al. [15] fabricated dense KNN ceramics using spark plasma sintering with special focus on piezoelectric property enhancement. Similarly, Wang and Li [16] studied a correlation between density, phase transition and property enhancement in KNN-based piezoceramics.

5.1.3 ELECTRON MICROSCOPY

In electron microscopy, a beam of accelerated electrons is allowed to interact with materials of all types at spatial resolutions significantly higher than those possible with conventional optical microscopy. This interaction in general provides the detailed topology, morphology, composition and crystallographic structure of the specimen. The standard procedure followed in all electron microscopes (EMs) is: a beam of electrons generated from electron gun (source) is accelerated towards the sample using a positive electrical potential. With the help of magnetic lens and metal apertures, a thin, monochromatic, focused beam of electrons is formed and directed onto the sample. Interactions take place inside the irradiated sample, influencing the electron beam. These interactions and effects are detected and finally transformed into an image. Regardless of the types of electron microscopy, the aforementioned procedures are followed in all EMs. After interaction with the bulk specimens, the back-scattered electrons provide information about the atomic number, the secondary electrons about the topology, and the Augar electrons about the composition of the specimen. Similarly, after interaction with thin specimens, the direct beam transmitted through the specimen carry the information about the thickness, the elastically scattered electrons carry the information of crystallographic structure, inelastic scattered electrons carry the information about the composition and bonding of the elements of the specimen. The electron–specimen interactions are

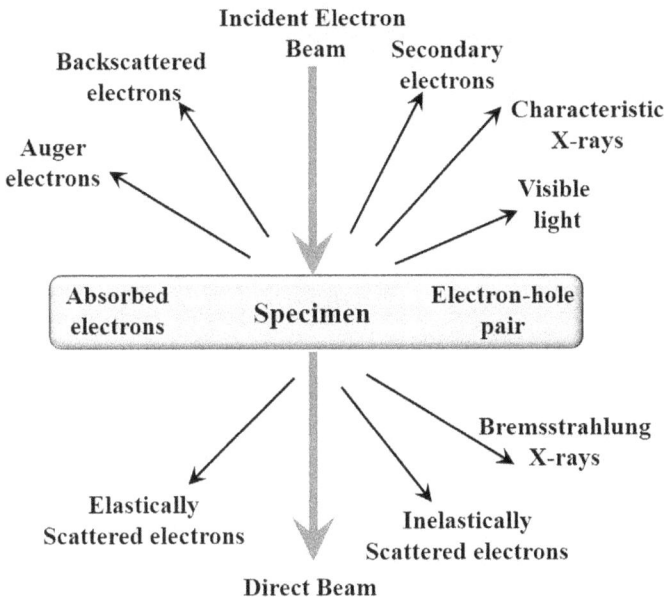

FIGURE 5.4 Schematic diagram illustrating electron-specimen interactions involved in electron microscopy.

schematically presented in Figure 5.4. In this section, a few important EMs which can be used to characterize the piezoelectric materials are discussed below.

5.1.3.1 Scanning Electron Microscopy

In piezoelectric materials, SEM is used to study the surface morphology, grain–grain boundary interface, the shape and size distribution of grains, solid–liquid pores and its distribution, composition, cross section morphology, fiber shape and size and so forth In a study made by Singha et al. [17] on $Na_{0.5}Bi_{0.5}TiO_3$ piezoceramics, the SEM micrographs clearly exhibit the change of microstructure with a change in sintering time from 1 hr to 24 hr. Initially, at 1 hr of sintering time, unimodal grain size distribution with a size of 2 µm was observed. However, on increasing the sintering time, the grain size increased and attained 5.2 µm at 24 hrs of sintering (Figure 5.5(a1-a2)). Interestingly, abnormal growth was also noticed in some grains on sintering the sample for 24 hrs. Control of microstructure plays a key role in regulating the electrical properties of polycrystalline ceramics. In this regard, scanning electron microscopy acts as an effective tool to study the change of microstructure. For instance, Wei et al. [18] in 2014 attempted to modulate the properties (specifically the electrical properties) of textured $Sr_{1.85}Ca_{0.15}NaNbO_{15}$ (SCNN) ceramics fabricated by reactive template grain growth (RTGG). The synthesis method itself signifies the grain growth along a particular axis and the authors demonstrated improved electrical properties due to texturing through alignment of the ceramic grains along the c-axis (the preferred

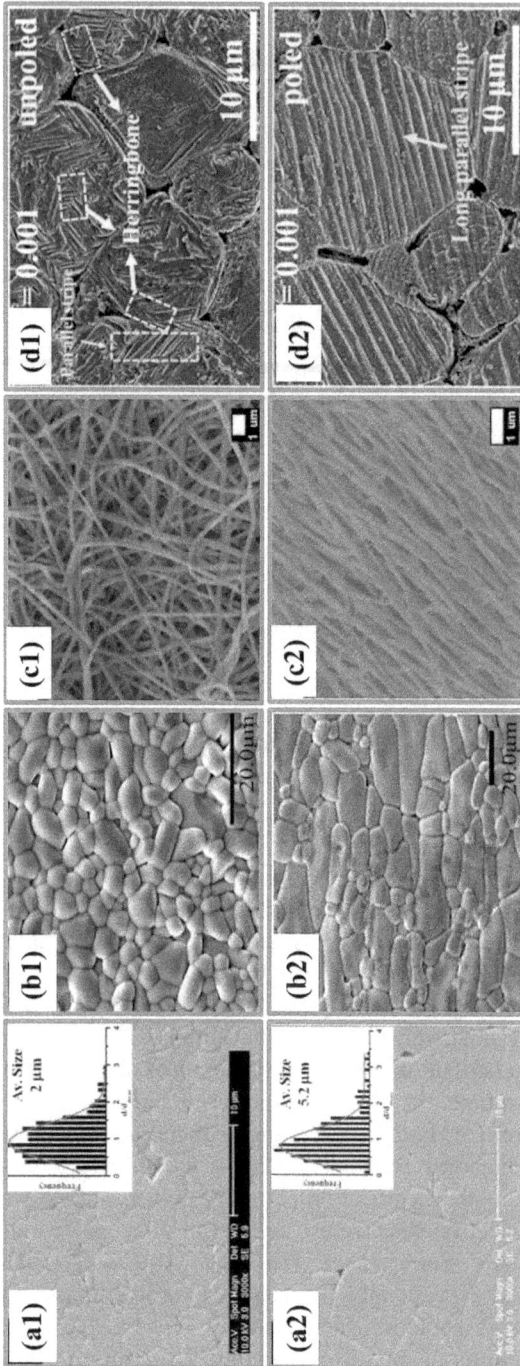

FIGURE 5.5   SEM micrographs showing the increase in grain size of NBT ceramic with increase in sintering time from (a1)1 hr to (a2) 24 hr; SEM pictures displaying grain orientation in the morphology of textured SCNN ceramics at sintering temperatures (b1) 1340°C for 4 hour and (b2) 1340°C for 6 hour; microstructure of (c1) electrospun and (c2) calcined KNN-2 NF (NF is mentioned in reference [19]); domain structures of 0.7BFMx-0.3BT (d1) before poling and (d2) after poling for x=0.001.

Source: [17, 18, 19, 20].

axis). Figure 5.5(b1-b2) shows an obvious difference in the grain orientation of SCNN ceramics grown by conventional mixed oxide route and RTGG method. In RTGG method, initially equiaxed grains with smaller size were more in number than anisotropic shape. However, with the increase in holding time, the textured anisotropic grains increased in number and size indicating improvement in texture quality. Recently, a study was made to investigate the potential of Li and Ta modified KNN piezoceramics as vibrational energy harvesters. In this report, it was noticed that there is strong co-relation between the microstructure and piezoelectric property of the nanofibers. Figure 5.5(c1-c2) displays a change in morphology of the nanofibers both before and after calcination (at a calcination temperature, 700 °C). At higher ramp rates, there is a loss of nanofiber morphology owing to porosity and crack formation because of rapid decomposition of binder material. The diameter of the fibers also reduced significantly post calcination and concomitant densification led to coarsened microstructure due to crystallite growth [19]. Besides, SEM remains a powerful method in initial investigations on the domain morphology and orientation. The pattern of unpoled $0.7BFM_{0.001}$–$0.3BT$ ceramic in Figure 5.5(d1) clearly shows randomly arrayed short parallel stripes and herringbone domains, however after poling the domains are turned into regular pattern (Figure 5.5(d2)) under an external electric field [20]. Zhou et al. [21] recently, examined the piezoelectric domain topographies of ceramics $0.96(K_{0.48}Na_{0.52})(Nb_{0.96}Sb_{0.04})O_3$-$0.04(Bi_{0.50}Na_{0.50})ZrO_3$ at different temperatures using temperature variable acid etching method (Figure 5.6(a, a2 and a3)). Poled samples were etched at three different temperatures (i.e.–60 °C, 25 °C and 80 °C) and it was observed that etching temperature had a profound influence on the domain morphology of the piezoceramics. Unique banded nano domains, wedge shaped domains and some long parallel domains were observed at–60 °C. The content of nanodomains increased with the increase of temperature and herring bone type of domains were finally visible at 25 °C. Parallel strips with small amount of nanodomains were visible at 80 °C. All these orientations indicate co-existence of different structures which is very important from the point of view of piezoelectric property. Nevertheless, morphology and size of mono-dispersed $BaZrO_3$ nanoparticles synthesized by seed assisted hydrothermal method were studied by Kanie and his group **[22]**. The results indicate highly uniform and spherical morphology of the nanoparticles with size of the particles gradually decreasing with increase in seeding amount in the precursor mixture. This is a significant study to estimate the regulated growth of particle prepared by hydrothermal method (Figure 5.6(b1, b2 and b3)). Besides, a change in the grain morphology during single crystal growth is very crucial and can be well investigated from SEM analysis. Moon et al. in 2011 [23] examined the grain coarsening behavior of a lead-free system $95Na_{1/2}Bi_{1/2}TiO_3$-$5BaTiO_3$ with addition of TiO_2 at different concentration (Figure 5.6(c1, c2 and c3)). Figure 5.6(c1-c3) clearly shows the three-dimensional grain shape with rounded cube edge which further changed with the amount of TiO_2. The shape change was expected to affect the coarsening behavior since the growth kinetics of rounded edged grains is influenced by the growth of facet planes.

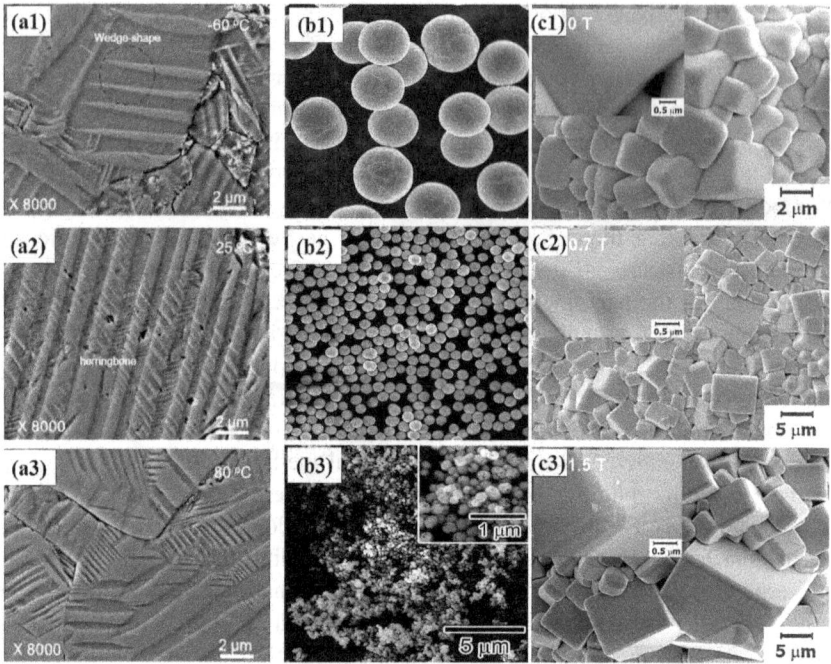

FIGURE 5.6 (a1), (a2) and (a3) SEM micrographs of domain topographies of poled KNNS-4%BNZ etched at -60°C, 25°C and 80°C respectively; (b1), (b2) and (b3) represent the FESEM micrographs of monodispersed $BaZrO_3$ particles synthesized by seed mediated hydrothermal process for Zs/Z_T =2.7 x 10^{-4}, 2.7 x 10^{-3} and 1.3 x 10^{-1} (mol/mol) respectively (Zs/Z_T indicate initial molar concentrations of Zr_{4+} ions supplied from seed and Zr-TEOA complex respectively); (c1), (c2) and (c3) present the SEM pictures of undoped, 0.7 mol% and 1.5 mol% TiO_2 doped 0.95NBT-0.05BT annealed at 1200°C for 10 hrs.

Source: [21, 22, 23].

5.1.3.2 Transmission Electron Microscopy

For piezoelectric material research, TEM is an extremely effective tool to study the nanoparticles, crystal structure, dislocations, grain boundaries, domain structure, chemical analysis, growth of layers and their compositions, defects, and so forth. A beam of energetic electrons allows to transmit through the thin specimen and the electron-atom interactions can be used to monitor the above features. High resolution TEM can be used to analyze the quality, shape, size, and density of quantum wells, wires, and dots. However, the process of sample preparation in TEM might be difficult as specimens must be less than 100 nm thick, which is equivalent to the mean free path of the electrons passing through them. The method used to prepare TEM specimens (e.g., ion milling, ion etching, chemical etching, mechanical milling, sample staining, etc.) depends on the material being analyzed and the kind of information that will be gleaned from the specimen. For most bulk piezoelectric materials, typically the sequence of preparation is ultrasonic disk cutting, dimpling, and ion-milling. Whereas for powders, microscopic organisms, nanotubes, nanofibers or viruses can be prepared by depositing a diluted sample containing the specimen onto films on support grids. Maurya et al. [24] while investigating hysteresis behavior of grain oriented $K_{0.5}Bi_{0.5}TiO_3$-$BaTiO_3$-$Na_{0.5}$-$Bi_{0.5}TiO_3$ (KBT-BT-NBT), employed high resolution TEM and SEM to demonstrate interface effect (Figure 5.7(a1-a2)). The low magnification image illustrates defect free interface that is attributed to lattice mismatch between KBT-BT-NBT textured grain and the BT seed templates. Such defect-free morphology gives rise to enhanced piezoelectric effect. The growth of textured grain incorporating dissolution and precipitation mechanism is evident from the TEM micrograph. Again in 2021, Maurya and his group [25] described a microscopic model based on high resolution TEM and neutron diffraction studies to explain the reasons behind high piezoelectric performance of lead-free $0.93(Na_{0.5}Bi_{0.5})TiO_3$-$0.07BaTiO_3$ (near MPB composition). The bright field TEM images in Figure 5.7(b1) and 5.7(b2) show a multitude of domain structures and the micrographs demonstrate ferroelastic lamellar domains of widths ~100–200 nm with planer $\{100\}_c$ domain walls. The HR-TEM images exhibit lattice fringes across the domains. Figure 5.7(c1) illustrates herringbone type domain structure which when magnified shows thick lamellar domains with smaller domain structure within Figure 5.7(c2). The corresponding SAED patterns and superlattice reflections are shown in Figure 5.7(b3) and 5.7(c3). Further, HR-TEM on NBT-BT whiskers were studied in the same paper Figure 5.8(a). It shows the presence of nanotwins in the form of lattice fringes. The FFT patterns obtained from a twin free region (region 1 in Figure 5.8(a) and twinned region 2 in Figure 5.8(a), were indexed for monoclinic symmetry. Monoclinic crystal structure with nanotwins is crucial owing to high degrees of freedom for monoclinic structure to adapt local stress field. The interfacial effects of lead free ferroelectric $(BaTiO_3$-$Ba(Cu_{1/3}Nb_{2/3})O_3)$ on HfO_2/Si (100) were studied using TEM by Kundu et al. [26] in 2015. Growth of defects at the interface usually deteriorates the electrical properties of the material. To reduce the charge injection from the interfacial layer, a good interface is essential. Figure5.8(b) shows cross-sectional high resolution TEM images of BT-BCN/HfO_2/Si stacks. On the other hand, interface defects can be observed in Figure 5.8(c) which depict crystalline quality with smooth, sharp interface of stacks and chemical stability between HfO_2 and BT-BCN layers. No secondary phases or inter diffusion were found in $BaTiO_3$-$Ba(Cu_{1/3}Nb_{2/3})O_3$ (BT-BCN) layers. Similarly, stripes of ferroelectric nanodomains can be found in Figure 5.8(d). Those nano-sized domains with very high mobility result in high domain wall density. This happens to be one of the principal reasons for increased piezoelectric response of the sample.

5.1.3.3 Atomic Force Microscopy

AFM is a potent technology that can image nearly every surface type, including glass, ceramics, composites, polymers, and biological samples. This technique is

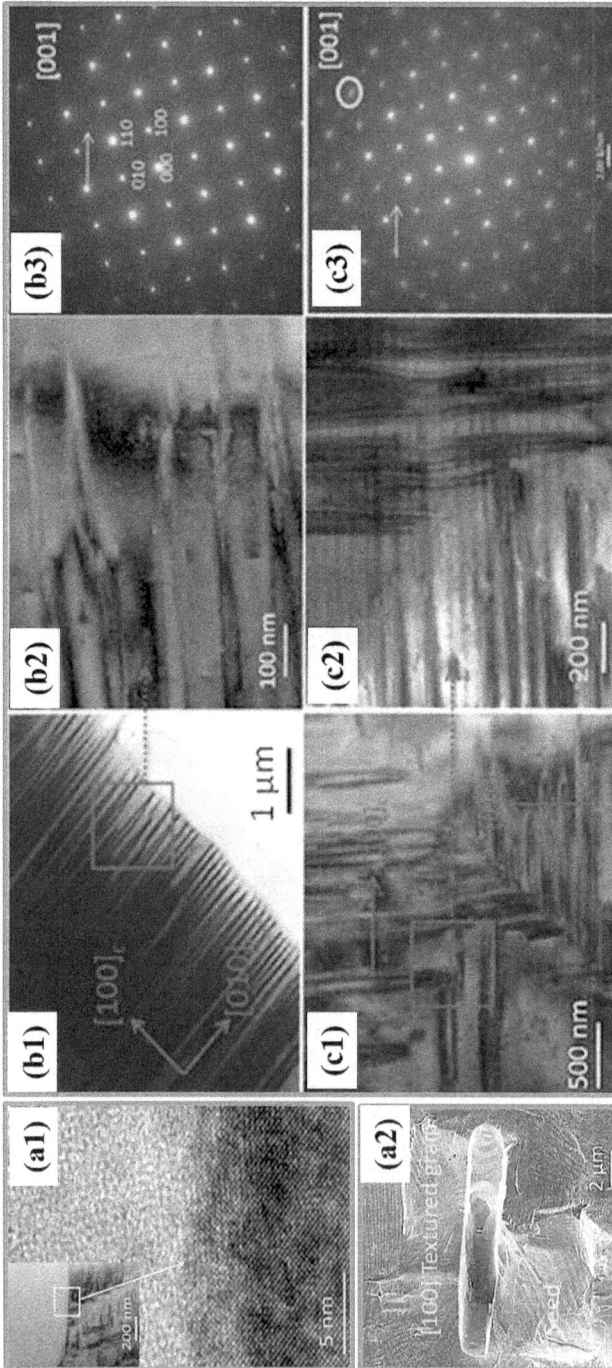

FIGURE 5.7 (a1) and (a2) HR-TEM image of BT seed interface with textured grain and SEM micrograph of textured of KBT-BT-NBT respectively; (b1) TEM image of thin domain structures with domain walls on (100) plane whose magnified view is shown in (b2); (c1) herringbone type domains consisting of thin domains with walls along (100) and (010) planes whose magnified view is shown in (c2); (b3) SEAD patterns and (c3) electron diffraction with nanobeam. All these refer to high purity crystal studies.

Source: [24, 25].

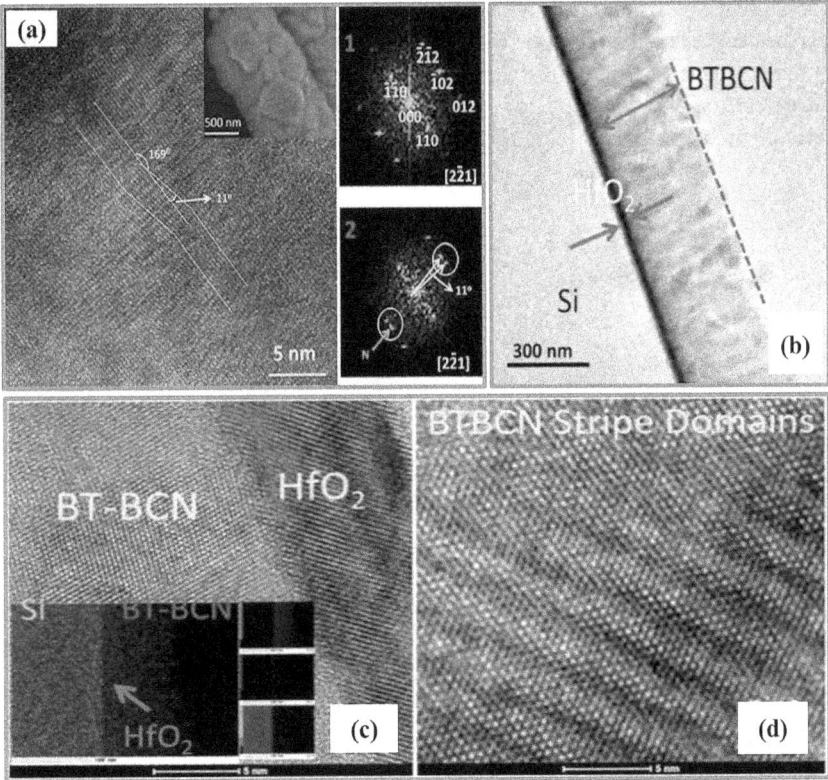

FIGURE 5.8 (a) HR-TEM micrograph indicating the presence of nanotwins in NBT-BT whiskers (patterns labeled as 1 and 2 represent the FFT patterns from the corresponding regions of HR-TEM image; yellow circular markings in FFT 2 shows atomic plane twinning and the inset to HR-TEM image presents the SEM micrograph of NBT-BT whisker); to illustrate the interface studies using HR-TEM, (b) shows the cross section TEM micrograph of BT-BCN on HfO2buffered Si; (c) HRTEM image of BT-BCN and HfO_2 interface (inset shows the elemental mapping of via high resolution EDS-TEM); (d) HRTEM pictures showing stripe nanodomains of BT-BCN.

Source: [25, 26].

best suited for characterizing nanoparticles and nanomaterials in air and liquid surroundings. Apparently, a wide range of particles varying from 1 nm to 8 μm can be characterized in the same scan. It primarily operates in contact mode and tapping mode. In the tapping modes, the AFM cantilever is vibrated above the sample surface in such a way that the tip is only occasionally in contact with the surface. In the contact modes, the AFM tip is continuously in contact with the surface. Moreover, high resolution and visualization in 3D images are provided by AFM from the movement of the tip. AFM provides qualitative as well as quantitative information about several physical properties like size, morphology, surface roughness and texture. It can also provide some statistical information such as size, volume distribution and surface area. For instance, nanoscale structure of piezoelectric polyvinylidene difluride (PVDF) was studied by Lee et al. [27] atomic force microscopy (AFM). The AFM images in Figure 5.9(a1) and (a2) depict that

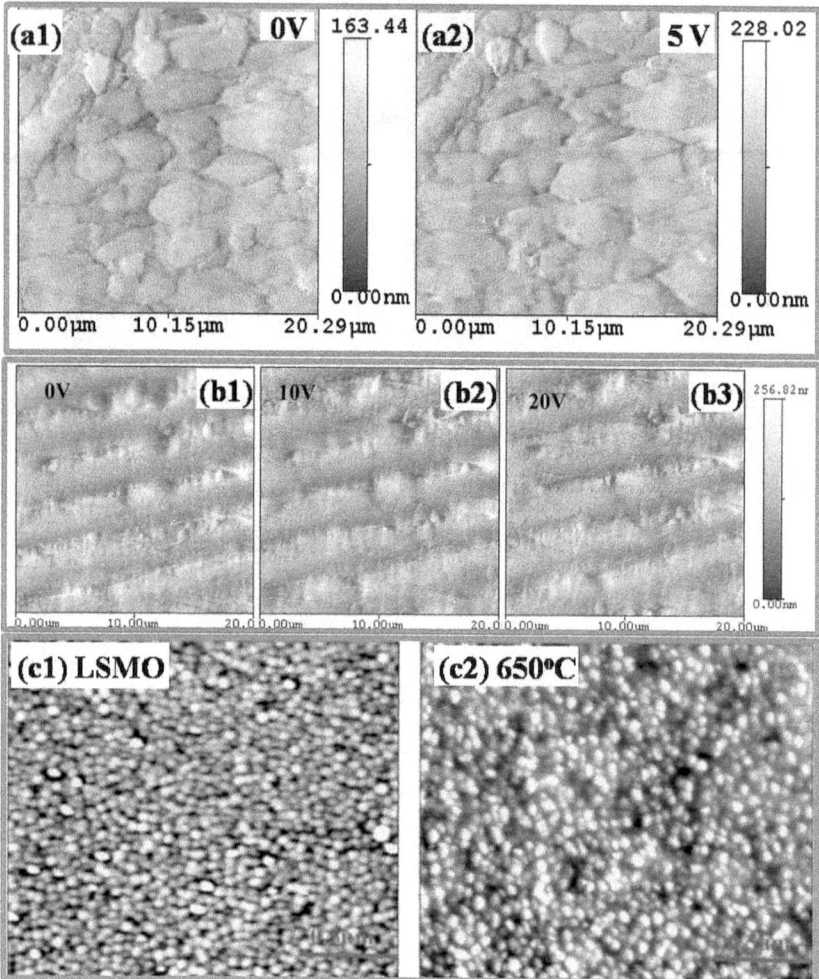

FIGURE 5.9 AFM images of change in surface texture of PVDF on application of electric field (a1) 0 V and (a2) 5 V; (b1), (b2) and (b3)AFM pictures of rate of surface change with applied electric fields of 0 V, 10 V and 20 V respectively; (c1) and (c2) represent the AFM morphology of LSMO bottom electrode and BCZT thin film at 650°C respectively.

Source: [27, 28].

the surface texture is squeezed at an applied electric field of 5 V. This was evident from the dipole realignment leading to increased surface height. With an increase in voltage, the peak to peak distance (D_{pp}) of the wavy surface altered itself signifying that D_{pp} decreased with an increase in applied potential at a rate 28.7 nm/V (Figure 5.9(b1)-(b3)). The domains become more aligned and can be seen in the morphology change [27]. Nevertheless, the surface morphology of lead-free $BaZr_{0.2}Ti_{0.8}O_3$-$Ba_{0.7}Ca_{0.3}TiO_3$ (BCZT) thin films grown on $La_{0.7}Sr_{0.3}MnO_3$ buffered Si(001) were investigated by Luo et al. [28] in 2013. With a grain size of up to 40 nm and a dense, smooth surface, the LSMO layer serves as an excellent template for the development of films. Next, it is discovered that when the deposition temperature of the BCZT films rises, so does the grain size (Figure 5.9(c1) and (c2)). AFM was employed as a useful tool for the immediate imaging of non-homogeneous electromechanical processes and the direct calculation of the piezoelectric constant at the local level. The calculated value of piezoelectric constant exhibited a wide distribution over the sample surface depicting the presence of non-homogeneous electro-mechanical processes [29]. AFM can be considered as an useful tool to observe nanofiber morphology, particularly in case of nanofibrous composites. In this regard, surface of the polybenzoxazole (PBO)/graphene added PVDF nanofibers was studied by Barstugan et al. [30]. The surface structure changed as per the type of PBO. The samples coded as P-D and P-E (sample code is as per the paper) exhibit least holes, humps and prongs on the surface leading to high piezoelectric effect. Further the surface roughness of BCZT films was estimated using AFM technique in a recent report. In a chosen area of 5 x 5 m², the average value of surface roughness was discovered to be 4.22 nm, demonstrating that the thin layer produced on the template was uniform and crack-free [31].

5.1.3.4 Piezoresponse Force Microscopy (PFM)

PFM is one of the voltage modulated AFM that enables imaging and manipulation of piezoelectric/ferroelectric domains at nanometer scale. It also enables the direct measurement of local physical characteristics (i.e., piezoelectric coefficients, nucleation bias, energy dissipation, disorder potential, and domain wall dynamics). It is done by putting a sharp and conductive probe in contact with a ferroelectric or piezoelectric material surface, and then biasing the probe with alternating current to cause the sample to deform due to the reverse piezoelectric effect. Standard split photodiode detector techniques are used to detect the consequent deflection of the probe cantilever, and a lock-in amplifier is then used to demodulate the signal. Following this method, high resolution images of both topography and ferroelectric domains can be acquired concurrently. Studying the local domain structure and how it changes with the applied electric field is crucial for understanding the static and dynamic features of ferroelectrics. In many cases, the formation of ferroelectric domains or phase influences the retention property of non-volatile memory devices. In such cases, PFM acts as a useful instrument. 180° domain wall movement between the c+ (positive polarizations, where the surface is positive) and c−(negative polarizations, where the surface is negative) domains was seen in the PFM topography taken on the surface of BT-BCN. The dark regions in the micrograph indicated polarization downwards while the polarization downwards was denoted by yellow confirming 180° phase difference between the domain configurations [26]. In a similar investigation using PFM, coexistence of ferroelectric and relaxor phases were confirmed after observing the color contrast between the brightest and dark regions [32]. Furthermore, Shi et al. [33] investigated the piezoelectric properties of $Ca_3Co_4O_9$ microplates by in situ PFM as shown in Figure 5.10. The microplates exhibited dense and homogeneous surface of the laminated structures along with large microplates. To excite the piezoelectric vibrations in the investigated sample, DC voltage of (5 V) is applied both in positive and negative directions. Interestingly, a clear contrast was observed in the micrographs indicating the existence of domain boundary. Furthermore, a strong piezoelectric response of the microplates was correlated with a sharp contrast in the amplitude image. Similar type of investigation was also made in a recent article where the authors revealed nanoscale domains with piezoelectric response [34].

5.1.4 Raman Spectroscopy

Raman Spectroscopy is a special method for observing rotational, vibrational, and other low-frequency modes in a system. It gives enough data to understand the local disorders, distortions, and strain present in the system since the vibrational spectrum has a characteristic shorter wavelength than that necessary for diffraction. Raman effect in crystals (solids) mainly takes into account the phonon vibrations rather than molecular. The first derivative of the polarizability with respect to the vibrational normal coordinate must be non-zero for a phonon to be Raman active. This can happen only in case of crystalline solids with no center of symmetry. In most of the piezoelectric materials, studies on the dynamics of structure (relaxor) have been carried out by analyzing the characteristic modes associated with nano regions; the selection rule being very sensitive to the local and global symmetries. Usually, Raman spectrum is presented as intensity versus difference in wave numbers between the scattered and incident beams and hence the obtained peaks are in agreement with the frequency of the phonon. Raman shift is normally obtained as:

$$\Delta\omega = \left(\frac{1}{\lambda_0} - \frac{1}{\lambda_1} \right),$$

Where, $\Delta\omega$= the Raman shift expressed in terms of wave number, 'λ_0' is the excitation or incident wavelength and 'λ_1' is the scattered wavelength. Since the wave vectors of the optical phonons are small, those participating in Raman Scattering of crystalline solids possess (conservation principle of wave vectors) lesser momentum in comparison to the Brillouin Zone. Hence only the zone centered phonons are involved in the scattering process. Based on this principle, Raman spectroscopic technique has been used as important characterization

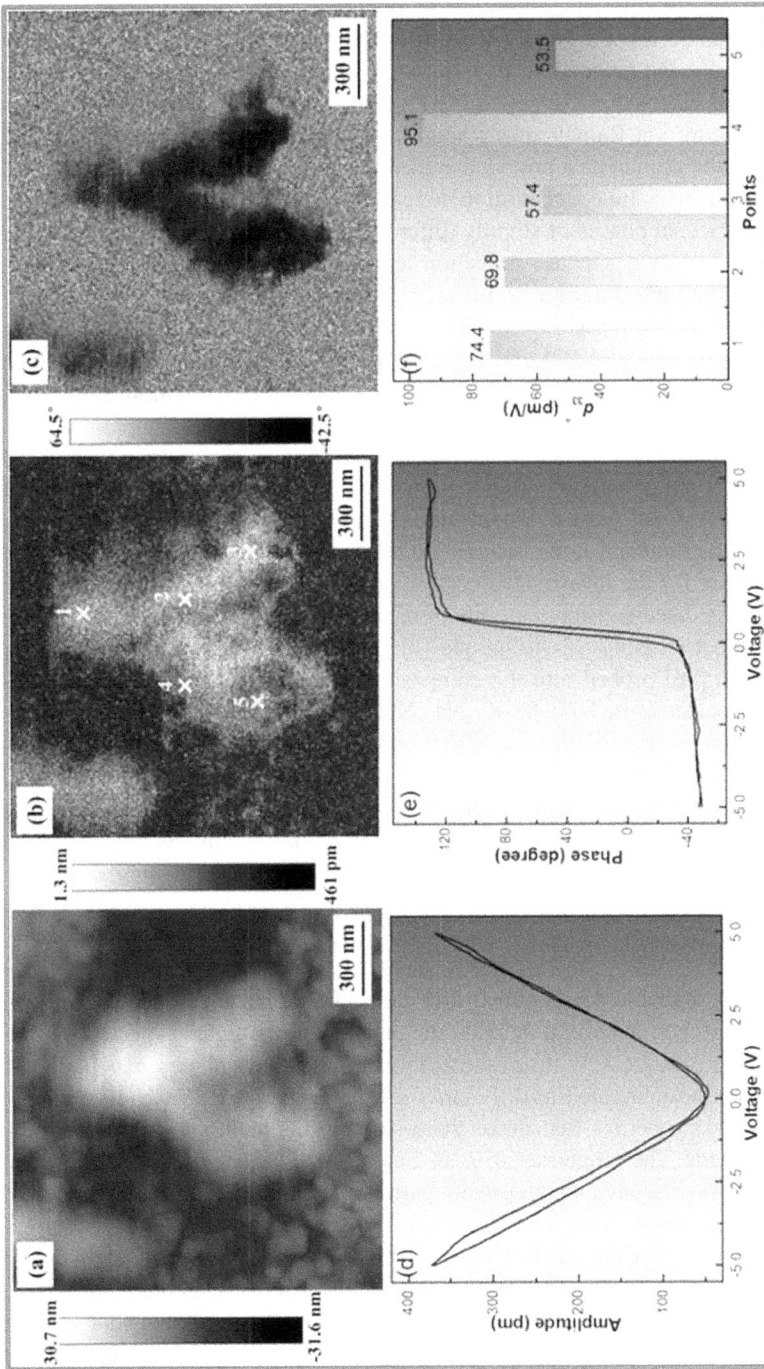

FIGURE 5.10 In situ PFM images of unpoled $Ca_3Co_4O_9$ microplates (a) topographic; (b) amplitude; (c) phase image; (d) amplitude-voltage butterfly loops; (e) phase-voltage hysteresis loops; (f) values of d_{33}^*.

Source: [33].

technique for structural investigation of materials. This popular and common technique has been used vividly used by many material science groups among which Dr. Rout and his teammates have done notable work. They have examined the structural changes in Ba doped NBT ceramics as a function of composition and temperature via Raman spectroscopy and XRD techniques. The anomalies observed in the Raman data provided evidence for the rhombohedral to tetragonal structural change as Ba concentration increased across the MPB. It also suggested that two tetragonal phases of slightly different space groups coexisted at elevated temperatures while the phase transition temperatures shifted towards left with increase in the concentration of Ba [35]. Previously, they also made a similar kind of phase transition study for NBT-SrTiO$_3$ (NBT-ST) system using Raman spectroscopy along with XRD, and dielectric analysis. Raman spectral investigation revealed the presence of MPB around 20 percent SrTiO$_3$ concentration. Further, two anomalies were clearly observable for 10 percent Sr at high temperatures via temperature dependent Raman spectroscopy. These anomalies correspond to the rhombohedral-tetragonal and tetragonal-cubic phase coexistence at 400 °C and 500 °C respectively. Finally, the phase transition temperatures derived from Raman and XRD were instrumental in constructing temperature and composition dependent phase diagram for the investigated ceramic system [4]. The same group has also systematically discussed about morphotropic phase boundaries in NBT-BT, a lead-free piezoelectric using this technique [6]. Following all these works, recently Singha et al. [36] probed into the compositional crossover from ferroelectric to relaxor ferroelectric in NBT-ST-K$_{0.5}$Na$_{0.5}$NbO$_3$ (NBT-ST-KNN) ceramics where Raman spectroscopic method is extensively analyzed for structural analysis (Figure 5.11). An extra mode round 150 cm^{-1} depicted the presence of nano sized domain structures. Additionally, triplet splitting of the band corresponding to the vibrations of TiO$_6$ octahedra and doublet splitting of the Ti^{4+}/Nb^{5+}–O band provided indirect proof of the co-existence of tetragonal and rhombohedral phases.

5.1.5 SURFACE AREA ANALYSIS

In view of sustainable development and global needs, nanotechnology has seen exponential growth in several fields in the past few years and is expected to have booming future progress. With advancement in synthesis techniques, nanoparticles/materials can be fabricated in different size, shape and structure leading to new and improved properties that make them suitable for numerous applications in almost all fields. The properties arise as they are small in size (1 to 100 nm) and have a wide surface area, having many particles per unit mass as compared with microparticles. Typically, the surface area measurement of materials is carried out using Brunauer–Emmett–Teller (BET) method. Since it is widely available in high purity and interacts significantly with the majority of materials, nitrogen is typically used in BET surface area analysis. Due to the normally poor contact between gaseous and solid phases, the surface must be cooled using liquid N$_2$ in order to produce significant amounts of adsorption. The sample cell is then filled

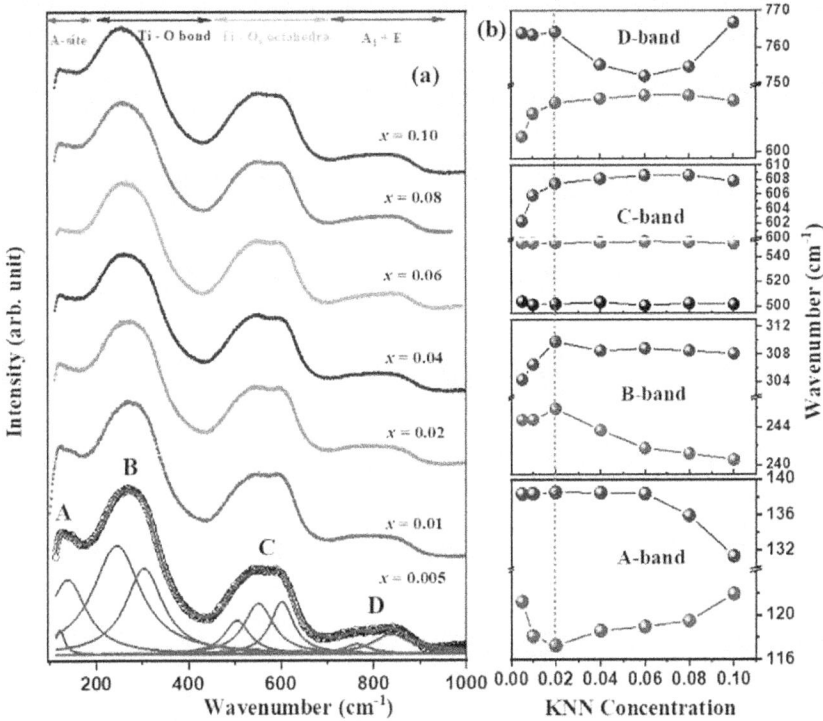

FIGURE 5.11 (a) Raman spectra at room temperature in the range 100-1000 cm⁻¹ and (b) compositional variation of wavenumber of NBT-ST-xKNN ($0.005 \leq x \leq 0.1$).

Source: [36].

with known quantities of nitrogen gas. By establishing circumstances of partial vacuum, it is possible to attain relative pressures lower than atmospheric pressure. No matter how high the pressure is raised, no more adsorption can occur after the saturation pressure. Pressure transducers that measure pressure changes with extreme accuracy and precision are used to track pressure changes brought on by the adsorption process. Once the adsorption layers have formed, the sample is removed from the nitrogen atmosphere, which allows the adsorbed nitrogen to be released from the substance and quantified. The gathered information is displayed using a BET isotherm, which depicts the amount of gas adsorbed as a function of relative pressure. For analysis of the data, the BET equation used is:

$$\frac{1}{v\left[1-\left(\frac{p}{p_o}\right)\right]} = \frac{c-1}{cv_m}\left(\frac{p}{p_o}\right) + \frac{1}{cv_m}$$

where p = the equilibrium pressure and p_o = saturation pressure of the adsorbates at adsorption temperature, c = the BET constant, v = the amount of gas adsorbed and v_m = the monolayer adsorbed gas quantity. Depending on the relative pressure ratio of p and p_o, the adsorption isotherms are of five types as shown in the schematic diagram (Figure 5.12). BET surface area analysis was used as one of the characterization techniques while optimizing the hydrothermal synthesis conditions of Nd doped BCTH ($[(Ba_{0.85}Ca_{0.15})_{0.995}Nd_{0.005}](Ti_{0.9}Hf_{0.1})O_3$) nano powders by He et al. [37]. They found that the adsorption-desorption curves of the samples (Figure 5.13(a)) were

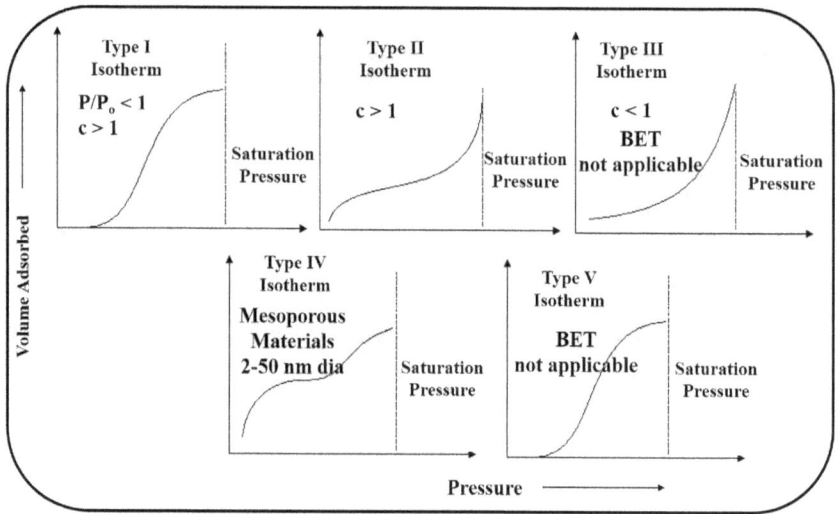

FIGURE 5.12 Schematic diagram displaying five types of adsorption isotherm based on relative pressure ratio (P/P_0).

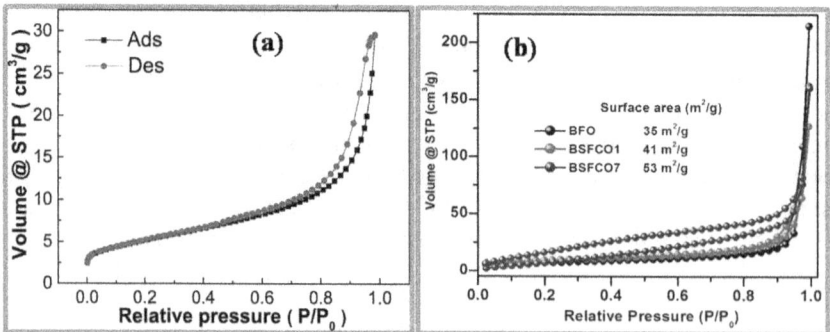

FIGURE 5.13 Adsorption and desorption curve of (a) Nd-BCTH powder; (b) BFO, BSFCO1 and BSFCO7.

Source: [37, 39].

S-shaped wrt P/P_0 axis where P = the equilibrium pressure of gas adsorption and P_0 = the saturated vapor pressure of the gas at the adsorption temperature. Again, the isotherm and its hysteresis were identified to be of type II and H_3 type respectively. The specific area calculated from this curve was 18.0207 m^2/g. In another study, the surface area and pore volume analysis were conducted for estimating the piezocatalytic activity of lead free $BaTiO_3$–based materials since materials having more surface area demonstrate better catalytic activity [38]. Effective BET surface area was evaluated for multiferroic Co, Sr doped $BiFeO_3$ nanoparticles by Puhan et al. [39] as shown in Figure 5.13(b). Highest surface area was obtained for $Bi_{1-x}Sr_xFe_{1-y}Co_yO_3$ ($x = 0.02$; $y = 0.07$) which offered a large number of active sites. The pore size distribution exhibited a disorderliness in porosity. Similar studies were also conducted by Puhan et al. [40] and Sahu et al. [41].

5.1.6 PARTICLE SIZE ANALYSIS

Particle size and its distribution play a basic and vital role in the fabrication, property development, efficiency enhancement, application, and research of powder materials. Particle size and distribution are typically measured using particle size analyzers based on various technologies, such as high definition image processing, Brownian motion analysis, light scattering, and so forth. Because they make sample optical characterization very simple, analyzers based on light scattering are particularly popular in academic research and many sectors. Additionally, in recent years, outstanding developments in light scattering technologies for particle characterization have attracted the interest of academics all around the world. Dynamic light scattering is one such light scattering method in which a laser beam is directed to pass through particles in suspension and the angular variation of the scattered light intensity is recorded. When compared to a laser beam, microscopic particles scatter light at large angles while large particles typically do so at small angles. The scattered intensity is then analyzed to determine the particle size using the Mie light scattering theory and the size of the particle is presented as a volume equivalent sphere diameter. Another scattering method called nanoparticle tracking analysis has recently come into existence. It uses image recording to detect individual particle movement during scattering. Using a nanoparticle size analyzer, the size of hydrothermally synthesized Nd-BCTH was analyzed under different conditions (Figure 5.14) [37]. However, the authors of this work stated that the nanoparticle size analyzer measured the size of agglomerates rather than single particle owing to the agglomeration of nanopowders. Based on the particle size, the suitable hydrothermal conditions for optimizing the sample (with least agglomeration) were selected. Further, to know the particle agglomeration and homogeneity during solid state synthesis of piezoelectric $KNN-Bi_{0.5}K_{0.5}TiO_3$ (KNN-BKT) ceramics, particle size estimation was carried out during the preparation stage by Pinheiro and Deivarajan [42]. The samples calcined at an optimized temperature of 900 °C were dispersed in ethanol medium for analysis. The fact that the powders were uniformly ground by high energy ball milling is demonstrated by a single modal particle size distribution and artefact peak. Although the average particle

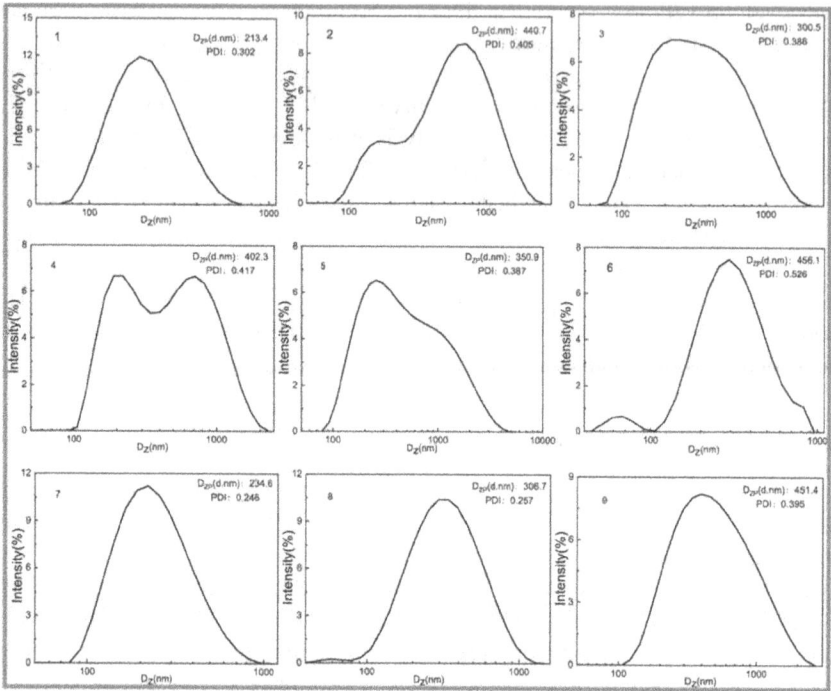

FIGURE 5.14 Particle size distribution of Nd-BCTH powders synthesized by hydrothermal process.

Source: [37].

size was 2.2 μm, it peaked at about 6.8 μm. A size distribution with such a wide range could result in good sample compaction in the end. Similar kind of study was also conducted previously to ensure homogeneous particle size during NBT synthesis by solid state reaction method. The particle sizes were checked before and after calcination and finally it was observed that as the duration of milling increased from 10 to 16 hrs, grain coarsening took place leading to degradation of piezoelectric properties [43].

5.2 THERMAL PROPERTIES

Thermal analysis is the process of measuring a material's physical and chemical characteristics in relation to temperature. The glass transition temperature, chemical decomposition, phase transition, phase diagrams, crystalline dynamics of glass and polymers, and the study of particular properties like enthalpy, heat capacity, coefficient of thermal expansion, and so forth may all be studied using thermal analysis techniques. In *differential thermal analysis*, the temperature gradient between the inert reference material and sample were recorded under identical heating or cooling cycles. The changes in the sample with respect to the inert reference can be found to be either exothermic or endothermic. *Thermogravimetric analysis* records the mass change of a sample in a particular atmosphere and provides information about evaporation, sorption and decomposition processes. Thermal analyses are typically used to examine how the initial precursors react to one another. Hence, Feizpour and his group [44] observed two pronounced exothermic peaks at 507 °C and 594 °C in the DTA curve during KNN synthesis (Figure 5.15). These peaks were in agreement with the negative peaks of the first TG derivative at 511 °C and 594 °C, indicating different stages of the reaction between alkali carbonates and Nb_2O_5, respectively. In another study, DTA/TGA analysis was conducted during different stages of preparation of $Ba_{0.85}Ca_{0.15}Zr_{0.1}Ti_{0.9}O_3$ lead free piezoelectric to confirm the various on-going chemical reactions. Weight loss of the powders in the temperature range 300–400 °C signifies decomposition of acetate ligands. Similar other decomposition peaks were also observed indicating different sol-gel reactions [45].

Similarly, change in the dimension of a specimen when subjected to a defined load is measured by *thermomechanical analysis* and it gives the information on

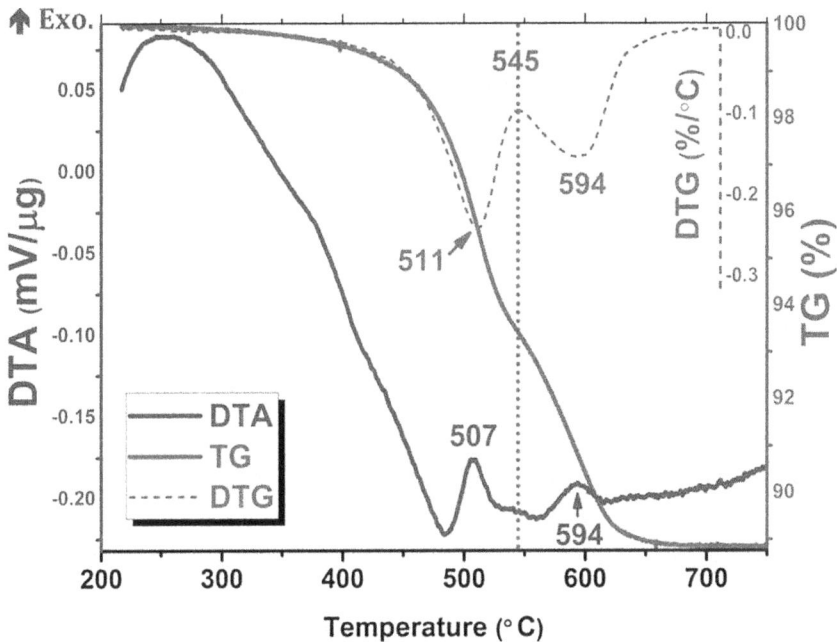

FIGURE 5.15   DTA, TG and DTG curves homogenized mixture of KNN precursors where temperatures of the peaks are written in °C.

Source: [44].

damping behavior, compliance, and mechanical module. The differential scanning calorimetry is a versatile technique to study phase transitions and associated parameters such as enthalpy change and entropy change at the phase transition of the piezoceramic samples. The sample and the reference material are kept at the same temperature throughout the heating process in this procedure. The additional heat input needed to keep the sample's temperature stable during any thermal event is measured and displayed against time or temperature. Depending on whether the change is endothermic or exothermic, the sample temperature either follows or lags the reference temperature during the thermal event. The specific heat is a characteristic property of the material and other parameters like enthalpy change, entropy can be calculated from it. Temperature dependent heat capacity (C_p) of Fe doped lead ytterbium was measured by Rout et al. [46]. Pure lead ytterbium exhibits a sharp peak around the antiferroelectric-ferroelectric phase transition of first order with significant value of entropy (Figure 5.16). On the other hand, the anomaly in heat capacity becomes broadened and weak with increasing Fe concentration. Heat and entropy of transition also start decreasing with increase in Fe concentration. On the same note, DSC heating and cooling curves in the temperature range 25–500 °C were studied by Fisher et al. [47] in KNN ceramics sintered in different atmospheres. The peaks indicate phase transitions were clearly visible in both oxygen and nitrogen atmosphere. However, tetragonal–cubic peak in the sample sintered in nitrogen atmosphere is small indicating non-attainment of equilibrium vacancy concentration during sintering. Similarly, other phase transitions were also explored in the light of DSC analysis in different sintering atmospheres.

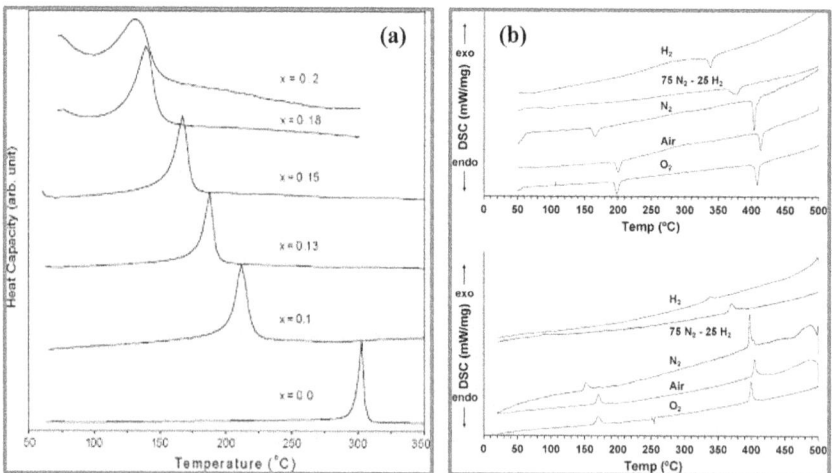

FIGURE 5.16   (a) Spectra showing the heat capacity of PFYT ceramics; (b) DSC heating and cooling curves of KNN ceramics sintered at 1100°C in different atmospheres such as O_2, air, N_2, $75N_2$-$25H_2$ and H_2.

Source: [46, 47].

5.3   ELECTROMECHANICAL PROPERTIES

5.3.1   DIELECTRIC

When a dielectric material is subjected to an external electric field, some non-zero macroscopic dipole moment is induced in it showing that it has been polarized under the influence of the applied field. The local electric field (E) and the induced dipole moment (P) are connected by the formula: P = αE, where α is the polarizability of the involved atom or molecule. In case of dielectric materials, polarization occurs due to the following possible mechanisms: (a) deformation polarization: it may be divided into electronic and atomic polarization. Generally, electronic polarization takes place when an externally applied electric field tends to displace the nuclei and the electrons in an atom. Due to the smaller mass of the electrons as compared to the nuclei, they respond rapidly to the changes in electric field (even at optical frequencies). Whereas, atomic polarization normally occurs when atoms or groups of atoms are displaced under the influence of an externally applied electric field. (b) orientation polarization: the applied electric field tends to form permanent dipoles and orient them in its direction. However, the thermal motion of the molecules may prevent this orientation (rotation). As a result, both temperature and the frequency of the applied electric field have a significant impact on the orientational polarization. (c) Ionic polarization: In an ionic lattice, the positive ions are redirected in the same direction as the applied field, while the negative ions are sent in the opposite direction, creating a dipole moment in the solid as a result. The nature of the interface at which the ions can concentrate is what mostly determines the ionic polarization, which exhibits only a very modest dependency on temperature. Hence, it relates to many cooperative processes in heterogeneous systems. The various types of polarization and their frequency range are presented in Figure 5.17(a). The macroscopic behavior of a dielectric material can be understood by using it between the plates of a parallel plate capacitor as shown in Figure 5.17(b). The capacitance (C) of the capacitor with a dielectric as medium is given by

$$C = \frac{\varepsilon_r \varepsilon_o A}{d},$$

where $\varepsilon_r = \varepsilon / \varepsilon_o$ is the relative permittivity and ε = the permittivity of the dielectric. ε_o (8.85×10^{-14} F/cm) = the permittivity of free space; d = thickness of the plates, and A = area of the plates. On applying an alternating voltage $V = V_o e^{j\omega t}$, the total current in the dielectric material is

$$I = j\omega \varepsilon_r C_o V,$$

where 'C_o' is the capacitance of vacuum and since ε_r is a complex quantity i.e. $\varepsilon_r = \varepsilon_r' - j\varepsilon_r''$, therefore I can be re-written as:

$$I = j\omega \left(\varepsilon_r' - j\varepsilon_r'' \right) C_o V$$

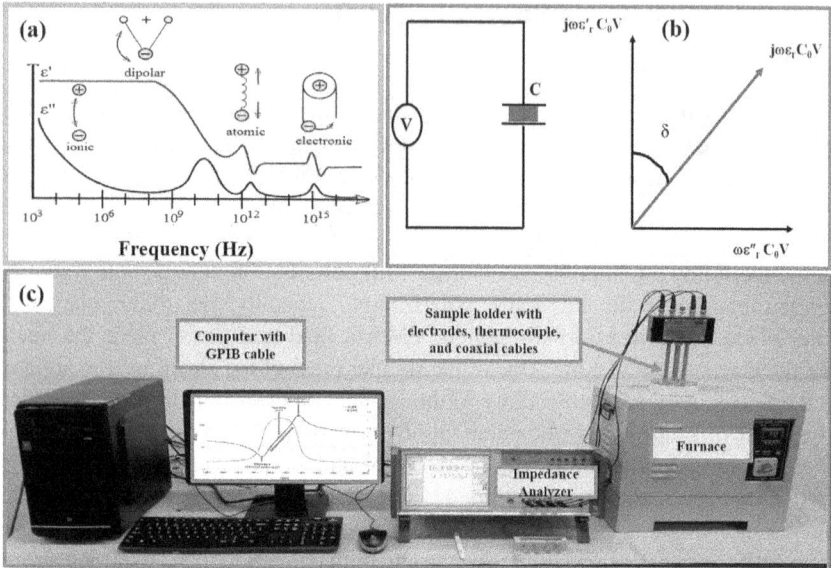

FIGURE 5.17 Schematic of (a) different types of polarization and their frequency range; (b) parallel plate arrangement for studying dielectric behavior of materials; (c) experimental set up of an impedance analyzer or LCR meter.

The value of ε_r' determines the out of phase element and ε_r'' determines the in phase element of current respectively.

Some part of the AC electrical energy is absorbed and dissipated by means of heat by the dielectric which is termed as the dielectric loss of the material. When the frequency of the applied field is in the range of the relaxation frequency of the dielectric, a condition of resonance is achieved. Under such circumstance, the current leads the voltage by (90-δ), where δ is the loss angle and tan δ signifies the electrical loss due to resonance and named as the loss tangent. The dielectric loss tangent is expressed as: $\tan \delta = \dfrac{\varepsilon_r''}{\varepsilon_r'}$

Generally, the measurements were carried out using an Impedance Analyzer or LCR meter. The detailed experimental set up is shown in Figure 5.17(c). The sintered pellets were electroded with silver paint by firing at 500 °C for 1 hr. The capacitance and loss tangent were measured in a wide frequency and temperature range. Using this technique, abundant experiments has been conducted by various researchers to study the dielectric behavior of the lead free ceramics. Our research group also works in this regard. In one of our investigations in 2010 [4], the temperature and composition dependent XRD, dielectric and Raman scattering studies were conducted to explore the phase transitions in $(1-x)Na_{0.5}Bi_{0.5}TiO_3-xSrTiO_3$, $0 \leq x \leq$

40. Ferroelectric-antiferroelectric-paraelectric phase transitions were thought to be the cause of anomalies in the dielectric constant that were detected as a function of temperature between 220–320 °C. On the other hand, the first anomaly in the dielectric loss versus temperature (200 °C) refers to the depolarization temperature. These temperatures shift towards the left (low temperatures) as the Sr concentration increases. However, the samples exhibited strong relaxor behavior with increase in Sr content. Finally, a temperature–composition phase diagram was proposed (Figure 5.18). In 2016, the dielectric behavior of a nearly piezoelectric system, that is $(0.8-x)(Na_{0.5}Bi_{0.5})TiO_3$-$0.2SrTiO_3$-$xBaTiO_3$ $(0.00 \le x \le 0.10)$ was examined by Praharaj et al. [48]. The diffuse phase transition behavior and large frequency dispersion of T_m (temperature corresponding to maximum dielectric constant) increases with the rise in barium titanate concentration. Further, the dielectric curve affirms strong relaxor nature of the samples as the $BaTiO_3$ content increases. Maximum dielectric constant was recorded for $x = 0.04$ (morphotropic phase

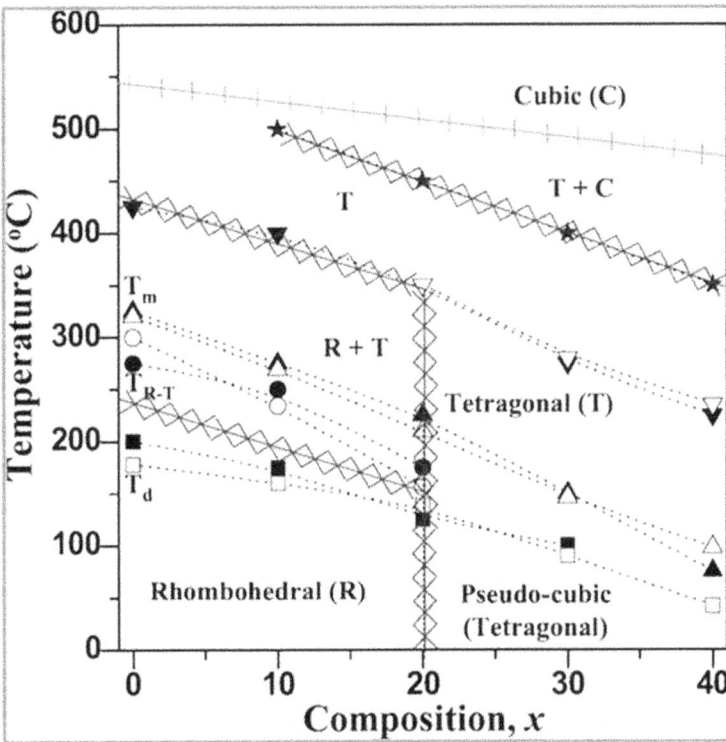

FIGURE 5.18 Temperature-composition phase diagram of (1-x)NBT-xST derived from dielectric (open symbols) and temperature dependent Raman spectroscopy (solid symbols) measurements.

Source: [54].

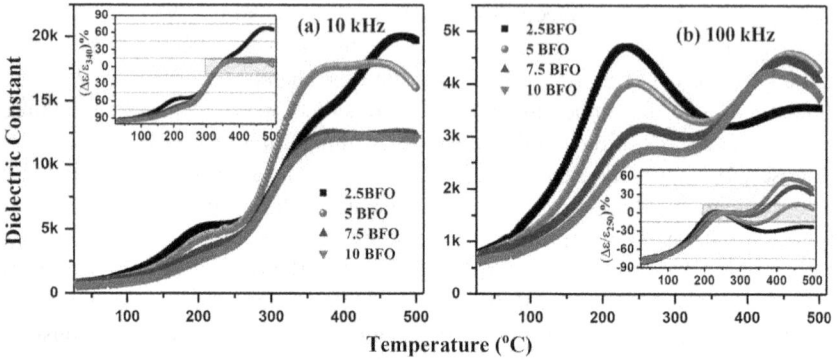

FIGURE 5.19 Temperature dependence of dielectric constant for different BFO samples at (a) 10 kHz and (b) 100 kHz. (inset: corresponding variation of dielectric constant within ±15%).

Source: [50].

boundary composition) which exhibited maximum piezoelectric property [49]. Moving further, they have also studied the effect of $BiFeO_3$ on the relaxor NBT-ST system [50]. This ceramic system is particularly important for its high-temperature stability of dielectric constant. In addition, it exhibits a TCε of −194 ppm/°C, which is perfect for various applications (Figure 5.19). Similar types of response are also examined in case of other NBT-ST based ternary systems [36]. In all these studies, the diffuseness of the dielectric constant increases with composition indicating high temperature stability of dielectric and indirectly piezoelectric properties.

5.3.2 P-E AND S-E LOOP

A flip in the polarization state combined with piezoelectric strain is one of the most often employed functional responses of the piezoelectric material. The P-E loop defines many reference points as per IEEE standard 180 like Ec, Pr, Ps and so forth as already mentioned in the introduction chapter. These parameters are essential from the point of view of the material manufacturers giving them indications of poling conditions for the effective use of the ceramic and a better perception of the material property. On the other hand, measurement of S-E loops is very important for actuator applications. One of the useful factors for developing actuators is the piezoelectric co-efficient d_{33}^*, which is given by the slope of the S-E loop. A proper quantification of the hysteresis in the output is highly essential for accurate positioning applications. Any actuator cannot return to its zero position in the presence of high degrees of hysteresis without a bipolar electric field drive rather than a unipolar one. Similarly, the onset of colossal non-linearity in the strain behavior with applied electric field can be measured with the help of S-E loops and can be illustrated with the appearance of the 'butterfly loop' (Figure 5.20 (c)).

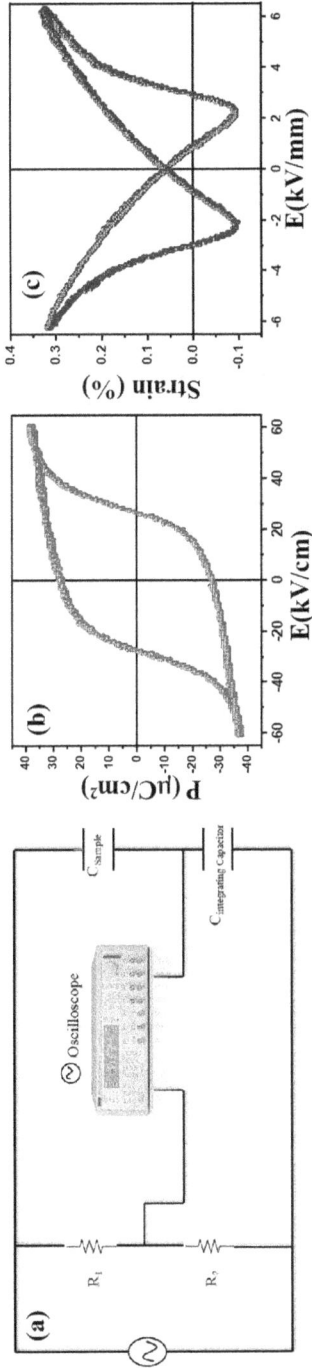

FIGURE 5.20 Schematic diagram showing (a) Sawyer-Tower circuit; (b) and (c) typical P-E loop of a ferroelectric material and S-E loop of a piezoelectric material respectively.

A schematic Sawyer-Tower circuit is shown in Figure 5.20(a). Polarization variations with electric field are useful in predicting the ferroelectric/ piezoelectric behavior of the samples. Usually, the ferroelectrics exhibit a square P-E loop (Figure 5.20 (b)) which may gradually change to a pinched/slim loop with the introduction of different phases such as antiferroelectric or relaxor phase in a compositional series. Co-existence of such phases in a compound give rise to large electric field induced strain. In a study by Praharaj et al. [49], such a gradual crossover was noticed in the P-E loops of $(0.8-x)(Na_{0.5}Bi_{0.5})TiO_3$-$0.2SrTiO_3$-$xBaTiO_3$ ($0.00 \leq x \leq 0.10$) (NBT-ST-BT). Square loop for $x = 0.00$ changed to very slim loop for $x = 0.10$ (Figure 5.21(a)). The composition ($x = 0.04$) around the existence of MPB (ferroelectric and relaxor) exhibited abnormalities in P_r and E_c values. Any actuator cannot return to its zero position in the presence of high degrees of hysteresis without a bipolar electric field drive rather than a unipolar one. (Figure 5.21(b)). Further, anomaly in remnant polarization and coercive field at this composition was explained on the basis of destabilization of long range ferroelectric order and establishment of relaxor nature. On examining bipolar strain versus electric field loop, a transformation was observed from butterfly shape ($x = 0.00$) to sprout shape ($x = 0.10$) (Figure 5.21(c)). A maximum strain (S_{max}) of 0.424% and d_{33}^* of 688 pm/

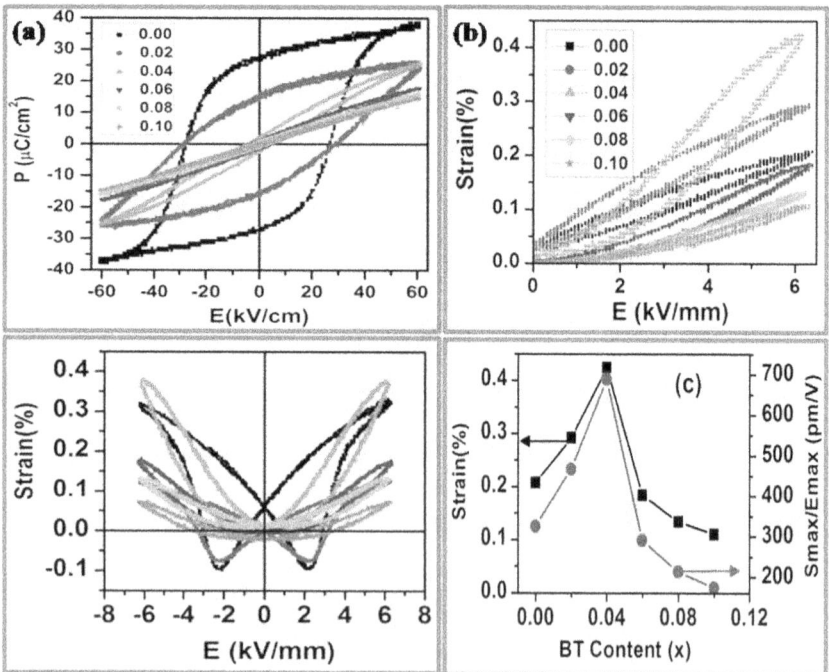

FIGURE 5.21 (a) P-E hysteresis loop; (b) unipolar S-E loops; (d) bipolar S-E loops and (c) Strain% and S_{max}/E_{max} for NBT-ST-xBT ceramics.

Source: [49].

V was noticed for $x = 0.04$ at 6 kV/mm (Figure 5.21(d)). Similar investigation was also made by our research group for $(0.8-x)(Na_{0.5}Bi_{0.5})TiO_3-0.2SrTiO_3-xBiFeO_3$ $(0.00 \leq x \leq 0.10)$ ceramics [50]. In another investigation, application of electric field on the P-E loops for $0.78Na_{0.5}Bi_{0.5}TiO_3-0.02SrTiO_3-0.02K_{0.5}Na_{0.5}NbO_3$ sample induced a peculiar pinching effect which was assigned to the pinning down of domain walls by defect dipoles. Domain wall motion and/or switching were discussed in detail with regard to the electric field dependence of hysteresis loops [36].

5.3.3 IMPEDANCE

Ac impedance spectroscopy (IS) is a strong and non-destructive experimental tool useful for characterizing electrical and dielectric relaxation of the polycrystalline compounds. The main advantages of the technique include automated simple electrical measurements, freedom of using arbitrary electrodes and co-relation of the results with properties such as composition, microstructural defects, dielectric relaxation behavior. Measurements can be taken over a broad spectrum of AC frequencies (1 mHz–1 MHz) depending upon the sophistication of the instrument. The obtained data can be interpreted in terms of different formalism: impedance Z^*, admittance Y^*, electric modulus M^*, and permittivity ε^*. Those formalism are interconnected as $Z^* = 1/Y^*$; $M^* = j\omega C_o Z^*$; $\varepsilon^* = 1/M^*$ and $\varepsilon^* = 1/ j\omega C_o Y^*$, where $j = \sqrt{-1}$, ω = angular frequency and $C_o = \varepsilon_o A/l$ (capacitance of the empty measuring cell), A = electrode area, and l being the distance between the electrodes, ε_o = dielectric permittivity of vacuum (8.854×10^{-14} Fcm^{-1}). Both A and l depend upon the sample dimensions. Heterogeneous dielectric materials representing various electroactive regions can be considered to be a parallel RC combination. The output response on applying a small AC signal to such materials, when plotted on a complex argand plane represents the contributions of grain, grain boundary and interface properties having separate time constants leading to three successive semicircles (depressed in non-ideal samples). The data may be displayed in the form of imaginary Z" (capacitive) against real Z′ (real) impedance. This plot is known as Nyquist plot. The relaxation time is obtained from a typical semicircle as $\tau = RC = 1/\omega_0$. In an ideal polycrystalline sample, the impedance plot exhibits a high frequency semicircular arc consistent with the inter-grain (boundary effect) and another low frequency semicircular arc reflecting intra-grain (grain effect) behavior along with a spike attributed by electrode or space charge effect. Figure 5.22 displays a typical impedance plot for an ideal polycrystalline compound.

In the previous section, 5.2.2, we have an impression that coexistence of phase boundaries such as ferroelectric and relaxor may be one of the important criteria for achieving good piezoelectric properties. In this regard, impedance spectroscopy has been used as a crucial tool by many researchers. This is because it provides indirect evidence of the existence of polar nano regions which are responsible for inducing relaxor behavior. In addition, it is also instrumental in explaining the diffuse nature of dielectric permittivity curve. Using this spectroscopic technique, different electroactive regions within a piezoelectric (basically dielectric) material can be segregated based on their relaxation time and temperature dependence.

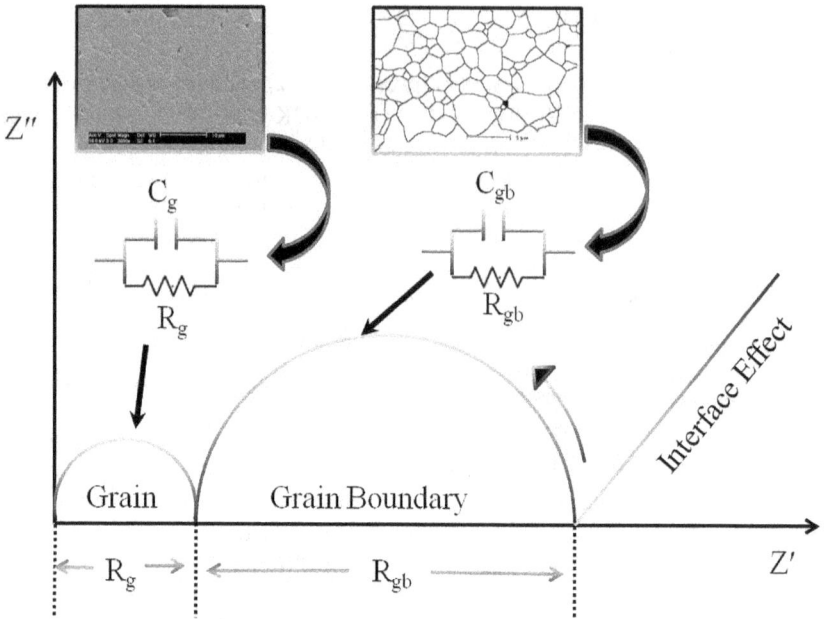

FIGURE 5.22 Typical impedance plot of an ideal polycrystalline compound showing grains and grain boundaries.

Apart from that, worthwhile information can be derived from the position of $M''_{max}(f)$ (imaginary part of M*) and $Z''_{max}(f)$ (imaginary part of Z*). Considering a parallel combination of single resistor (R) and capacitor (C) representing an electroactive region,

$$M'' = \frac{C_0}{C}\left[\frac{\omega RC}{1+(\omega RC)^2}\right].$$ (1)

$$Z'' = R\left[\frac{\omega RC}{1+(\omega RC)^2}\right]$$ (2)

Using the above equations (1) and (2), position of maxima in $M''(f)$ and $Z''(f)$ curve is given by,

$$\omega_{max} = 2\pi f_{max}$$ (3)

Applying equation (3) in equations (1) and (2), height of maxima i.e. M''_{max} and Z''_{max} can be calculated as

$$M''_{max} = \frac{C_0}{2C} = \frac{1}{2\varepsilon_r}. \tag{4}$$

$$Z''_{max} = \frac{R}{2} \tag{5}$$

After correction for sample geometry,

$$\frac{M''_{max}}{\varepsilon_0} = \frac{1}{2\varepsilon_0\varepsilon_r} = \frac{1}{2C_{corr}}. \tag{6}$$

Here, C_{corr} is the corrected capacitance, usually calculated as pellet thickness/area.

Moreover, the maxima in $M''(f)$ and $Z''(f)$ are dominated by constituents with smallest capacitance (grain response) and largest resistance (grain boundary response) respectively. Further, occurrence of $M''_{max}(f)$ and $Z''_{max}(f)$ at close frequencies suggest same electroactive region (with same time constant). In contrast, separate positions of maxima can be related to localized dielectric relaxation of resistive grain boundaries. Rodel and his group used this formulation to study the existence and response of the nanoscopic polarizable regions for the first time in 2014 [51, 52]. Zang et al. [51] discovered the presence of highly polarizable phase in addition to bulk response by analyzing $M''(f)$ spectra (shoulder region) of CaZrO$_3$ modified 0.94Bi$_{1/2}$Na$_{1/2}$TiO$_3$-0.06BaTiO$_3$ and 0.82(0.94BNT-0.06BT)-0.18(K$_{1/2}$Na$_{1/2}$)NbO$_3$ (KNN) ceramics. The relaxation frequency of these polar regions followed V-F law and Arrhenius law below and above the Burn's temperature 'T$_B$' respectively. Such evidences pointed towards the resemblance of these regions with polar nano regions (PNRs). Presence of highly polarizable phase was revealed in case of (1-x)(0.94BNT-0.06BT)-xKNN (x = 0, 0.03, 0.09, 0.18) using $M''(f)$ spectra. Additionally, they provided an explanation for the variation in temperature-dependent permittivity with composition in terms of thermal evolution of polar nano regions induced by KNN addition [52]. Another lead free piezoelectric, NBT-ST-BT [53] having dense microstructure was studied by IS over a wide range of temperatures (50–400 °C). The evolution of shoulder at high frequencies in the imaginary part of modulus spectra provided an indirect indication that the ceramic series contains highly polarizable elements (Figure 5.23). These polarizable entities might be the nano sized domain architectures within the grains (Figure 5.24) [53]. This is further supported by the study of relaxation dynamics of PNRs in 0.78NBT-0.2ST-0.02BT system via complex electric modulus formalism. The dynamics of PNRs across T$_B$, T$_m$ and T$_f$ were analyzed extensively [54].

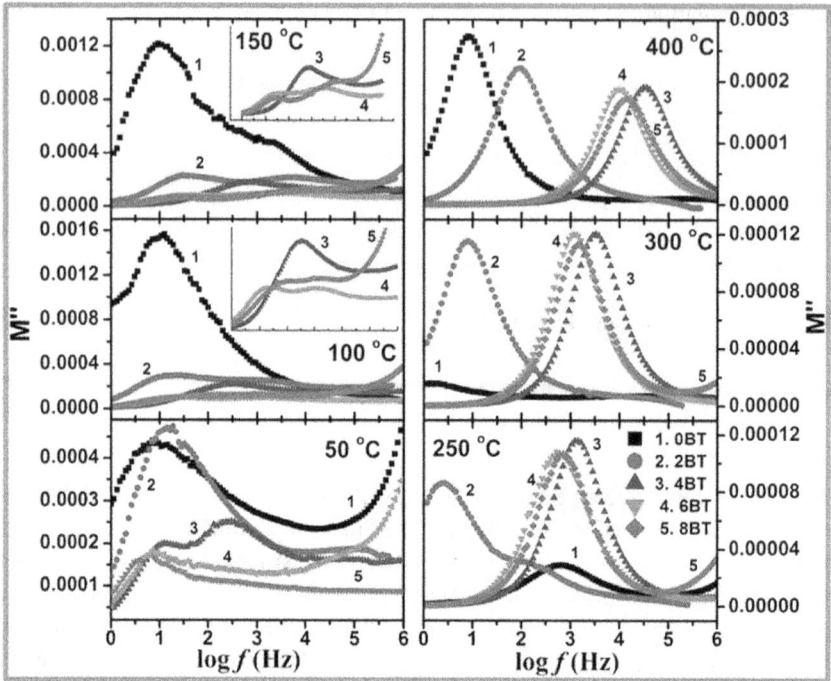

FIGURE 5.23 M" vs logf for NBT-ST-xBT (x=0.00≤x≤0.08) samples showing the existence of a shoulder in the high frequency range at different temperatures.

Source: [53].

5.3.4 DC RESISTIVITY

To fully comprehend the associated electromechanical properties of piezoelectric materials, it is crucial to take into account their electrical conductivity. For instance, electrical conductivity has a direct impact on the efficiency of the poling process, and breakdown strength is also highly correlated with conductivity. Since electrical conductivity is a substantially temperature-dependent feature, it is highly desirable to take it into account for high-temperature piezoelectric applications. Using a voltage step technique, the DC conductivity is measured by applying a DC voltage to an electroded sample and recording the charging current until a steady-state current can be produced. The true steady-state current, which is equal to the DC current, can be estimated by fitting the charging current with a power law. A Keithley Electrometer can be used to measure the current. The electrometer can serve as a source of voltage as well. The DC conductivity can also be derived from conductivity study using impedance spectroscopy. Usually, high DC resistivity is a basic requirement for poling materials with high Curie temperatures. In conjunction with the above fact, Yan et al. [55] studied the DC resistivity of perovskite like layered $Nd_2TiO_2O_7$ and La_2TiO_7 piezoelectric ceramics with possibly very high

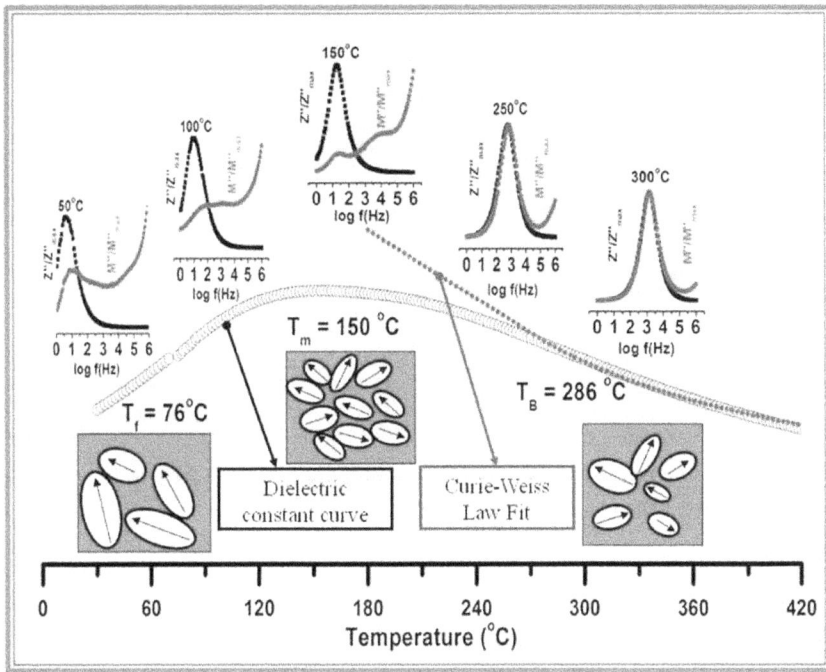

FIGURE 5.24 Schematic diagram representing the thermal evolution of PNRs in NBT-ST-0.08BT derived from the simultaneous plot of normalized Z" and M". It shows a crossover from long range to short range relaxation processes in the temperature range T_B to T_f.

Source: [53].

Curie temperatures. The DC resistivity of these ceramics was found to be lower in a direction perpendicular to the pressing direction as compared to the parallel direction. This anisotropy was lessened at higher temperatures. Further, resistivity of La_2TiO_7 nanometer powders was found to be more than 1 mΩ at 700°C, proving it to be useful for high temperature piezoelectric applications. Li et al. [56] also studied dc resistivity as a key figure of merit that substantially affects the high-temperature application of piezoelectric ceramics. They noticed an increase in DC resistivity with the percentage of Cu/Nb co-doping on $Bi_4Ti_3O_{12}$ (BTCN) piezoelectric ceramics in the temperature range 200–650 °C (Figure 5.25). A very high resistivity of 8.39×10^6 Ω.cm at 500 °C could be obtained for BTCN-1.5 due to induced oxygen vacancies by the co-doping. Such an increase in dc resistivity was explained in the light of impedance spectroscopy. Similar works were carried out by many other researchers to enhance the dc resistivity of piezoceramics [57-59].

5.3.5 d_{33}

As discussed earlier in chapter 2, the piezoelectric charge constant (d) is defined as the mechanical strain (S) experienced by a piezoelectric material per unit

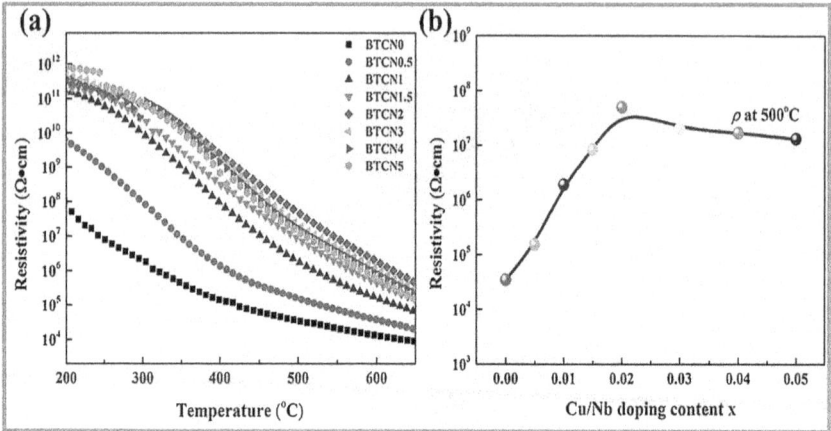

FIGURE 5.25 (a) DC resistivity of BTCN ceramics from 200 to 650°C; (b) DC resistivity as a function of Cu/Nb doping on BTCN ceramics.

Source: [56].

applied electric field, or alternatively, the polarization created per unit mechanical stress (T) applied to a piezoelectric material. The first subscript to d denotes the direction of the polarization that is produced in the material when the electric field, E, is zero or, alternately, the direction of the applied field intensity. The second subscript indicates either the direction of the produced strain or the direction of the applied tension. For example: in case of d_{33}, both induced polarization and applied stress are in direction 3 or induced strain and applied electric field are in direction 3, that is, parallel to the polarized direction. Since the strain inflicted on a piezoelectric material by an applied electric field is calculated as the product of the value for the electric field and the value for d; d is a key factor in determining a material's suitability for actuator (strain-dependent) applications. The d_{33} values of a piezoceramic can be measured by a simple and straightforward Berlincourt method associated with a quasi-static piezo d_{33}-meter. The ceramic sample is poled prior to the measurement, and the relative magnitudes of the charge output and the applied tiny oscillating force can then be quantified in comparison to a sample with a known and certified piezoelectric coefficient. Since piezoelectric materials frequently are used as actuators, the displacement of piezo-materials in presence of an electric field is a concern. The displacement, however, is too small to be quickly quantified using a standard technique. In this regard, optical interferometry provides the ability for the precise measurement of small displacements within units of nanometers due to its extraordinarily high resolution and lack of a need for length scale calibration. Additionally, optical interferometry is capable of measuring without using any mechanical contact. Laser interferometers that use a single or dual beam can detect piezoelectric coefficients. It can also be measured using an impedance analyzer and the resonance and anti-resonance method. To calculate the piezoelectric coefficients, it is essential to acquire the following parameters such as the resonant frequencies, capacitance, and density of the sample along with sample dimension. Measurement of d_{33} is a necessary condition for estimating the piezoelectric performance of any material. Especially, the ceramics posses higher values of d_{33} than other category of materials. In this regard, BT and KNN ceramics show encouraging results. In 2014, Zheng et al. [60] designed a unique combination of giant d_{33} and high T_C by doping different concentrations of $Bi_{0.5}Li_{0.5}ZrO_3$ (BLZ) in $K_{0.40}Nb_{0.60}Sb_{0.035}O_3$ ceramic. Co-existence of rhombohedral and tetragonal phases in the sample series was held responsible for achieving high level of d_{33} (400 pC/N for 3 mol% of BLZ) along with high T_C (292 °C for 3 mol% of BLZ). Similarly, high piezoelectric coefficient of 375 pC/N was also attained by $(Ba_{1-x}Ca_x)(Ti_{0.98}Zr_{0.02})O_3$ ceramics for $x = 0.01$ [61]. In another work, the effect of poling on the d_{33} values of textured (1-x-y)BNT-xBT-yKNN was studied. d_{33} was measured via direct effect using Pennebaker Model 8000 d_{33} meter and it was found that texturing could improve the piezoelectric response of the sample reaching a maximum of 245 pC/N [62].

5.3.6 g_{33}

The electric field that a piezoelectric material produces per unit of applied mechanical stress, or the mechanical strain that a piezoelectric material suffers per unit of applied electric displacement, is the piezoelectric voltage constant, or g. The initial subscript to g denotes the direction of the generated electric field or applied electric displacement in the material. The second subscript indicates either the direction of the produced strain or the direction of the applied tension; g is essential for determining a material's appropriateness for sensing (sensor) applications since the strength of the induced electric field produced by a piezoelectric material in response to an applied physical stress is the product of the value for the applied stress with g. Unlikely d_{33}, g_{33} suggests that both induced electric field and the stress are in direction 3 (parallel to polarized direction of the ceramic element) or both induced strain and the electric displacement are in direction 3. The g_{33} value was determined using the relation $g_{33} = d_{33} / (\varepsilon_0 \cdot \varepsilon_r)$. In an very recent and interesting work by Zhang [63], a metal-free small organic ferroelectric was developed having high piezoelectric voltage coefficients, $g_{33} \sim 437.2 \times 10^{-1}$ Vm/N and $g_{31} \sim 586.2 \times 10^{-3}$ Vm/N. He first measured the piezoelectric charge coefficient d_{33} using piezoforce microscopy (PFM). The thin film prepared by him was excited via PFM tip across the resonant frequency in the vertical PFM mode at 2V. On comparing the results with that of PVDF film at the same drive voltage, close resemblance in resonant frequency was noted; d_{33} was estimated as z-axis tip vibration in vertical PFM mode and thereafter, g_{33} was calculated. There are also previous reports on the measurement of voltage coefficient in the literature [64, 65].

5.4 SUMMARY

Characterization of the samples post synthesis plays a crucial role in estimating important properties of materials including piezoelectrics. Starting right from the basic structural and morphological investigations (such as XRD, Raman spectroscopy, density measurement, surface area analysis, SEM, TEM, DSC, DTA-TGA to characterizations specific for determination of piezoelectric figures of merit (e.g., P-E and S-E loop; d_{33} and g_{33}; dielectric and impedance spectroscopy) are necessary for evaluating the performance of the materials. With this motive, this current chapter describes all the important techniques for characterization piezoelectric materials.

REFERENCES

1. Jones, J.L., 2007. The use of diffraction in the characterization of piezoelectric materials. *Journal of Electroceramics*, 19(1), pp. 69–81.
2. Noheda, B., Cox, D.E., Shirane, G., Gonzalo, J.A., Cross, L.E. and Park, S.E., 1999. A monoclinic ferroelectric phase in the $Pb(Zr_{1-x}Ti_x)O_3$ solid solution. *Applied Physics Letters*, 74(14), pp. 2059–2061.
3. Guo, R., Cross, L.E., Park, S.E., Noheda, B., Cox, D.E. and Shirane, G., 2000. Origin of the high piezoelectric response in $Pb(Zr_{1-x}Ti_x)O_3$. *Physical Review Letters*, 84(23), p. 5423.
4. Kim, S., Choi, H., Han, S., Park, J.S., Lee, M.H., Song, T.K., Kim, M.H., Do, D. and Kim, W.J., 2017. A correlation between piezoelectric response and crystallographic structural parameter observed in lead-free $(1-x)(Bi_{0.5}Na_{0.5})TiO_3-xSrTiO_3$ piezoelectrics. *Journal of the European Ceramic Society*, 37(4), pp. 1379–1386.
5. Rout, D., Moon, K.S., Kang, S.J.L. and Kim, I.W., 2010. Dielectric and Raman scattering studies of phase transitions in the $(100-x)Na_{0.5}Bi_{0.5}TiO_3-xSrTiO_3$ system. *Journal of Applied Physics*, 108(8), p. 084102.
6. Rout, D., Moon, K.S., Rao, V.S. and Kang, S.J.L., 2009. Study of the morphotropic phase boundary in the lead-free $Na_{1/2}Bi_{1/2}TiO_3-BaTiO_3$ system by Raman spectroscopy. *Journal of the Ceramic Society of Japan*, 117(1367), pp. 797–800.
7. Habib, M., Lee, M.H., Choi, H.I., Kim, M.H., Kim, W.J. and Song, T.K., 2020. Phase evolution and origin of the high piezoelectric properties in lead-free $BiFeO_3-BaTiO_3$ ceramics. *Ceramics International*, 46(14), pp. 22239–22252.
8. Ochoa, D.A., Esteves, G., Iamsasri, T., Rubio-Marcos, F., Fernández, J.F., García, J.E. and Jones, J.L., 2016. Extensive domain wall contribution to strain in a (K, Na) NbO_3-based lead-free piezoceramics quantified from high energy X-ray diffraction. *Journal of the European Ceramic Society*, 36(10), pp. 2489–2494.
9. Moon, K.S., Rout, D., Lee, H.Y. and Kang, S.J.L., 2011. Solid state growth of $Na_{1/2}Bi_{1/2}TiO_3-BaTiO_3$ single crystals and their enhanced piezoelectric properties. *Journal of Crystal Growth*, 317(1), pp. 28–31.
10. Rout, D., Han, S.H., Moon, K.S., Kim, H.G., Cheon, C.I. and Kang, S.J.L., 2009. Low temperature hydrothermal epitaxy and Raman study of heteroepitaxial $BiFeO_3$ film. *Applied Physics Letters*, 95(12), p. 122509.
11. Rout, S.K., Hussain, A., Sinha, E., Ahn, C.W. and Kim, I.W., 2009. Electrical anisotropy in the hot-forged $CaBi_4Ti_4O_{15}$ ceramics. *Solid State Sciences*, 11(6), pp. 1144–1149.
12. Hussain, A., Ahn, C.W., Lee, H.J., Kim, I.W., Lee, J.S., Jeong, S.J. and Rout, S.K., 2010. Anisotropic electrical properties of $Bi_{0.5}(Na_{0.75}K_{0.25})_{0.5}TiO_3$ ceramics fabricated by reactive templated grain growth (RTGG). *Current Applied Physics*, 10(1), pp. 305–310.

13. Egerton, L. and Dillon, D.M., 1959. Piezoelectric and dielectric properties of ceramics in the system potassium—sodium niobate. *Journal of the American Ceramic Society*, 42(9), pp. 438–442.

14. Jaeger, R.E. and Egerton, L., 1962. Hot pressing of potassium-sodium niobates. *Journal of the American Ceramic Society*, 45(5), pp. 209–213.

15. Wang, K., Zhang, B.P., Li, J.F. and Zhang, L.M., 2008. Lead-free $Na_{0.5}K_{0.5}NbO_3$ piezoelectric ceramics fabricated by spark plasma sintering: Annealing effect on electrical properties. *Journal of Electroceramics*, 21(1), pp. 251–254.

16. Wang, K. and Li, J.F., 2012. (K,Na)NbO$_3$-based lead-free piezoceramics: Phase transition, sintering and property enhancement. *Journal of Advanced Ceramics*, 1(1), pp. 24–37.

17. Singha, A., Praharaj, S. and Rout, D., 2021. Effect of sintering time on microstructure and electrical properties of lead-free sodium bismuth titanate perovskite. *Materials Today: Proceedings*, 46, pp. 4568–4573.

18. Wei, L., Chao, X., Han, X. and Yang, Z., 2014. Structure and electrical properties of textured $Sr_{1.85}Ca_{0.15}NaNb_5O_{15}$ ceramics prepared by the reactive templated grain growth. *Materials Research Bulletin*, 52, pp. 65–69.

19. Ichangi, A., Shvartsman, V.V., Lupascu, D.C., Lê, K., Grosch, M., Schmidt-Verma, A.K., Bohr, C., Verma, A., Fischer, T. and Mathur, S., 2021. Li and Ta-modified KNN piezoceramic fibers for vibrational energy harvesters. *Journal of the European Ceramic Society*, 41(15), pp. 7662–7669.

20. Luo, F., Li, Z., Chen, J., Yan, Y., Zhang, D., Zhang, M. and Hao, Y., 2022. High piezoelectric properties in 0.7BiFeO$_3$–0.3BaTiO$_3$ ceramics with MnO and MnO$_2$ addition. *Journal of the European Ceramic Society*, 42(3), pp. 954–964.

21. Zhou, C. and Zhang, J., 2020. Feasible acid-etching method for investigating temperature-dependent domain configurations of ferroelectric ceramics. *Journal of the European Ceramic Society*, 40(13), pp. 4469–4474.

22. Kanie, K., Seino, Y., Matsubara, M. and Muramatsu, A., 2017. Size-controlled hydrothermal synthesis of monodispersed BaZrO$_3$ sphere particles by seeding. *Advanced Powder Technology*, 28(1), pp. 55–60.

23. Moon, K.S., Rout, D., Lee, H.Y. and Kang, S.J.L., 2011. Effect of TiO$_2$ addition on grain shape and grain coarsening behavior in 95Na$_{1/2}$Bi$_{1/2}$TiO$_3$–5BaTiO$_3$. *Journal of the European Ceramic Society*, 31(10), pp. 1915–1920.

24. Maurya, D., Zhou, Y., Wang, Y., Yan, Y., Li, J., Viehland, D. and Priya, S., 2015. Giant strain with ultra-low hysteresis and high temperature stability in grain oriented lead-free $K_{0.5}Bi_{0.5}TiO_3$-BaTiO$_3$-Na$_{0.5}$Bi$_{0.5}$TiO$_3$ piezoelectric materials. *Scientific Reports*, 5(1), pp. 1–8.

25. Maurya, D., Murayama, M., Pramanick, A., Reynolds Jr, W.T., An, K. and Priya, S., 2013. Origin of high piezoelectric response in A-site disordered morphotropic phase boundary composition of lead-free piezoelectric 0.93(Na$_{0.5}$Bi$_{0.5}$)TiO$_3$–0.07BaTiO$_3$. *Journal of Applied Physics*, 113(11), p. 114101.

26. Kundu, S., Maurya, D., Clavel, M., Zhou, Y., Halder, N.N., Hudait, M.K., Banerji, P. and Priya, S., 2015. Integration of lead-free ferroelectric on HfO$_2$/Si (100) for high performance non-volatile memory applications. *Scientific Reports*, 5(1), pp. 1–10.

27. Lee, H., Cooper, R., Wang, K. and Liang, H., 2008. Nano-scale characterization of a piezoelectric polymer (polyvinylidene difluoride, PVDF). *Sensors*, 8(11), pp. 7359–7368.

28. Luo, B.C., Wang, D.Y., Duan, M.M. and Li, S., 2013. Growth and characterization of lead-free piezoelectric $BaZr_{0.2}Ti_{0.8}O_3$–$Ba_{0.7}Ca_{0.3}TiO_3$ thin films on Si substrates. *Applied Surface Science*, 270, pp. 377–381.

29. Stamopoulos, D. and Zhang, S.J., 2014. A method based on optical and atomic force microscopes for instant imaging of non-homogeneous electro-mechanical processes and direct estimation of d_{ij} coefficients in piezoelectric materials at the local level. *Journal of Alloys and Compounds*, 612, pp. 34–41.

30. Barstugan, R., Barstugan, M. and Ozaytekin, I., 2019. PBO/graphene added β-PVDF piezoelectric composite nanofiber production. *Composites Part B: Engineering*, 158, pp. 141–148.

31. Liu, S., Zhang, Z., Shan, Y., Hong, Y., Farooqui, F., Lam, F.S., Liao, W.H., Wang, Z. and Yang, Z., 2021. A flexible and lead-free BCZT thin film nanogenerator for biocompatible energy harvesting. *Materials Chemistry Frontiers*, 5(12), pp. 4682–4689.

32. Dinh, T.H., Kang, J.K., Lee, J.S., Khansur, N.H., Daniels, J., Lee, H.Y., Yao, F.Z., Wang, K., Li, J.F., Han, H.S. and Jo, W., 2016. Nanoscale ferroelectric/relaxor composites: origin of large strain in lead–free Bi–based incipient piezoelectric ceramics. *Journal of the European Ceramic Society*, 36(14), pp. 3401–3407.

33. Shi, Z., Cao, S., Araújo, A.J., Zhang, P., Lou, Z., Qin, M., Xu, J. and Gao, F., 2021. Plate-like $Ca_3Co_4O_9$: A novel lead-free piezoelectric material. *Applied Surface Science*, 536, p. 147928.

34. Li, P., Huan, Y., Yang, W., Zhu, F., Li, X., Zhang, X., Shen, B. and Zhai, J., 2019. High-performance potassium-sodium niobate lead-free piezoelectric ceramics based on polymorphic phase boundary and crystallographic texture. *Acta Materialia*, 165, pp. 486–495.

35. Rout, D., Moon, K.S., Park, J. and Kang, S.J.L., 2013. High-temperature X-ray diffraction and Raman scattering studies of Ba-doped $(Na_{0.5}Bi_{0.5})TiO_3$ Pb-free piezoceramics. *Current Applied Physics*, 13(9), pp. 1988–1994.

36. Singha, A., Praharaj, S., Rout, S.K. and Rout, D., 2022. Composition dependent crossover from ferroelectric to relaxor-ferroelectric in NBT-ST-KNN ceramics. *Current Applied Physics*, 36, pp. 160–170.

37. He, X., Fang, B., Wu, D., Lu, X., Zhang, S. and Ding, J., 2022. Tailoring synthesis conditions for $[(Ba_{0.85}Ca_{0.15})_{0.995}Nd_{0.005}](Ti_{0.9}Hf_{0.1})O_3$ nanopowders by hydrothermal method and their luminescence properties. *Chemical Physics*, 562, p. 111675.

38. Li, J., Wei, X., Sun, X.X., Li, R., Wu, C., Liao, J., Zheng, T. and Wu, J., 2022. *A Novel Strategy for Excellent Piezocatalytic Activity in Lead-Free BaTiO$_3$-Based Materials via Manipulating the Multiphase Coexistence.* ACS Applied Materials & Interfaces.

39. Puhan, A., Bhushan, B., Satpathy, S., Meena, S.S., Nayak, A.K. and Rout, D., 2019. Facile single phase synthesis of Sr, Co co-doped $BiFeO_3$ nanoparticles for boosting photocatalytic and magnetic properties. *Applied Surface Science*, 493, pp. 593–604.

40. Puhan, A., Bhushan, B., Meena, S.S., Nayak, A.K. and Rout, D., 2021. Surface engineered Tb and Co co-doped $BiFeO_3$ nanoparticles for enhanced photocatalytic and magnetic properties. *Journal of Materials Science: Materials in Electronics*, 32(6), pp. 7956–7972.

41. Sahu, S.K., Zlotnik, S., Navrotsky, A. and Vilarinho, P.M., 2015. Thermodynamic stability of lead-free alkali niobate and tantalate perovskites. *Journal of Materials Chemistry* C, 3(29), pp. 7691–7698.

42. Pinheiro, E.D. and Deivarajan, T., 2019. Influence of porous configuration on dielectric and piezoelectric properties of KNN–BKT lead-free ceramic. *Applied Physics A*, 125(11), pp. 1–8.

43. Chou, C.S., Yang, R.Y., Chen, J.H. and Chou, S.W., 2010. The optimum conditions for preparing the lead-free piezoelectric ceramic of $Bi_{0.5}Na_{0.5}TiO_3$ using the Taguchi method. *Powder Technology*, 199(3), pp. 264–271.

44. Feizpour, M., Ebadzadeh, T. and Jenko, D., 2016. Synthesis and characterization of lead-free piezoelectric $(K_{0.50}Na_{0.50})NbO_3$ powder produced at lower calcination temperatures: A comparative study with a calcination temperature of 850 °C. *Journal of the European Ceramic Society*, 36(7), pp. 1595–1603.

45. Castkova, K., Maca, K., Cihlar, J., Hughes, H., Matousek, A., Tofel, P., Bai, Y. and Button, T.W., 2015. Chemical synthesis, sintering and piezoelectric properties of $Ba_{0.85}Ca_{0.15}Zr_{0.1}Ti_{0.9}O_3$ leadfree ceramics. *Journal of the American Ceramic Society*, 98(8), pp. 2373–2380.

46. Rout, D., Subramanian, V., Hariharan, K. and Sivasubramanian, V., 2006. Diffuse phase transition of Fe doped lead ytterbium tantalate ceramics. *Solid State Communications*, 137(8), pp. 446–450.

47. Fisher, J.G., Rout, D., Moon, K.S. and Kang, S.J.L., 2009. Structural changes in potassium sodium niobate ceramics sintered in different atmospheres. *Journal of Alloys and Compounds*, 479(1–2), pp. 467–472.

48. Praharaj, S., Rout, D., Subramanian, V. and Kang, S.J., 2016. Study of relaxor behavior in a lead-free $(Na_{0.5}Bi_{0.5})TiO_3$-$SrTiO_3$-$BaTiO_3$ ternary solid solution system. *Ceramics International*, 42(11), pp. 12663–12671.

49. Praharaj, S., Rout, D., Kang, S.J. and Kim, I.W., 2016. Large electric field induced strain in a new lead-free ternary $Na_{0.5}Bi_{0.5}TiO_3$-$SrTiO_3$-$BaTiO_3$ solid solution. *Materials Letters*, 184, pp. 197–199.

50. Praharaj, S., Singha, A. and Rout, D., 2021. Dielectric and piezoelectric properties of lead-free $Na_{0.5}Bi_{0.5}TiO_3$-$SrTiO_3$-$BiFeO_3$ ternary system. *Journal of Alloys and Compounds*, 867, p. 159114.

51. Zang, J., Li, M., Sinclair, D.C., Frömling, T., Jo, W. and Rödel, J., 2014. Impedance Spectroscopy of $(Bi_{1/2}Na_{1/2})TiO_3$–$BaTiO_3$ Based HighTemperature Dielectrics. *Journal of the American Ceramic Society*, 97(9), pp. 2825–2831.

52. Zang, J., Li, M., Sinclair, D.C., Jo, W. and Rödel, J., 2014. Impedance spectroscopy of $(Bi_{1/2}Na_{1/2})TiO_3$–$BaTiO_3$ ceramics modified with $(K_{0.5}Na_{0.5})NbO_3$. *Journal of the American Ceramic Society*, 97(5), pp. 1523–1529.

53. Praharaj, S., Rout, D., Anwar, S. and Subramanian, V., 2017. Polar nano regions in lead free $(Na_{0.5}Bi_{0.5})TiO_3$-$SrTiO_3$-$BaTiO_3$ relaxors: an impedance spectroscopic study. *Journal of Alloys and Compounds*, 706, pp. 502–510.

54. Praharaj, S., Subramanian, V., Kang, S.J. and Rout, D., 2018. Origin of relaxor behavior in $0.78(Na_{0.5}Bi_{0.5})TiO_3$–$0.2SrTiO_3$–$0.02BaTiO_3$ ceramic: An electrical modulus study. *Materials Research Bulletin*, 106, pp. 459–464.

55. Yan, H., Ning, H., Kan, Y., Wang, P. and Reece, M.J., 2009. Piezoelectric ceramics with super-high curie points. *Journal of the American Ceramic Society*, 92(10), pp. 2270–2275.

56. Li, X., Chen, Z., Sheng, L., Li, L., Bai, W., Wen, F., Zheng, P., Wu, W., Zheng, L. and Zhang, Y., 2019. Remarkable piezoelectric activity and high electrical resistivity in Cu/Nb co-doped $Bi_4Ti_3O_{12}$ high temperature piezoelectric ceramics. *Journal of the European Ceramic Society*, 39(6), pp. 2050–2057.

57. Chen, H., Shen, B., Xu, J. and Zhai, J., 2013. The grain size-dependent electrical properties of $Bi_4Ti_3O_{12}$ piezoelectric ceramics. *Journal of Alloys and Compounds*, *551*, pp. 92–97.

58. Sanson, A. and Whatmore, R.W., 2002. Properties of $Bi_4Ti_3O_{12}$–$(Na_{1/2}Bi_{1/2})TiO_3$ piezoelectric ceramics. *Japanese Journal of Applied Physics*, *41*(11S), p. 7127.

59. Chen, Z., Sheng, L., Li, X., Zheng, P., Bai, W., Li, L., Wen, F., Wu, W., Zheng, L. and Cui, J., 2019. Enhanced piezoelectric properties and electrical resistivity in W/Cr co-doped $CaBi_2Nb_2O_9$ high-temperature piezoelectric ceramics. *Ceramics International*, *45*(5), pp. 6004–6011.

60. Zheng, T., Wu, J., Cheng, X., Wang, X., Zhang, B., Xiao, D., Zhu, J., Lou, X. and Wang, X., 2014. New potassium–sodium niobate material system: a giant-d_{33} and high-TC lead-free piezoelectric. *Dalton Transactions*, *43*(30), pp. 11759–11766.

61. Li, W., Xu, Z., Chu, R., Fu, P. and Zang, G., 2010. High piezoelectric d_{33} coefficient in $(Ba_{1-x}Ca_x)(Ti_{0.98}Zr_{0.02})O_3$ lead-free ceramics with relative high Curie temperature. *Materials Letters*, *64*(21), pp. 2325–2327.

62. Fancher, C.M., Blendell, J.E. and Bowman, K.J., 2013. Poling effect on d_{33} in textured $Bi_{0.5}Na_{0.5}TiO_3$-based materials. *Scripta Materialia*, *68*(7), pp. 443–446.

63. Zhang, H.Y., 2022. A small-molecule organic ferroelectric with piezoelectric voltage coefficient larger than that of lead zirconate titanate and polyvinylidene difluoride. *Chemical Science*, *13*(17), pp. 5006–5013.

64. Sun, Y., Chang, Y., Wu, J., Liu, Y., Jin, L., Zhang, S., Yang, B. and Cao, W., 2019. Ultrahigh energy harvesting properties in textured lead-free piezoelectric composites. *Journal of Materials Chemistry A*, *7*(8), pp. 3603–3611.

65. Li, Z., Fan, X., Yi, J., Tan, S., Zhang, Z., Lu, T., Zhang, L. and Zhu, W., 2022. Outstanding Piezoelectric Sensitivity of Poly (Vinylidene-Trifluoroethylene) for Acceleration Sensor Application. *IEEE Transactions on Dielectrics and Electrical Insulation*.

6 Piezoelectric Energy Harvesting Structure and Mechanism

6.1 STRUCTURE

The rapid growth in the development of micro and nanoscale materials has led to major advancements in designing piezoelectric transducers. These transducers can harvest the complacent energy accessible in the ambient atmosphere of the electronic devices and utilize it to power them. The mechanical energy conversion can take place in three steps: (a) external excitation inducing mechanical vibrations (associated with piezo transducers' mechanical stability under impedance matching and high stress); (b) conversion of mechanical to electrical energy (correlated with electromechanical coupling factor); and (c) conversion of electrical energy into power delivered (depends on electrical impedance matching and design of the circuit). In the third step, the electrical energy from the high-impedance piezoelectric transducer is transferred to a rechargeable low-impedance battery using an appropriate DC/DC converter. Besides, piezoelectric transducers/energy harvesters (PEH) are expected to deliver maximum power under a certain mechanical load. Hence, selection of a suitable geometry and mode of operation is of utmost importance. The basic principle of operation of PEH involves the coupling of ambient vibrations with the piezo-structures, which produces alternating bending strains. These bending strains are then transformed into AC voltage by the piezoelectric material. The usual frequency generated by different vibration sources in our surroundings is listed in Table 6.1. In order to achieve maximum amplitude of the AC voltage, frequency matching seems to be an important parameter. The most common architecture employed in PEH is a cantilever beam since it has a high electromechanical coupling factor. Apart from that, cymbal and stacked structures are also popular. In the following sections, we discuss the different piezoelectric structures and modes of operation.

DOI: 10.1201/9781003317289-6

TABLE 6.1
Different vibration sources and their frequency of oscillation

Vibration Source	Frequency of oscillation (in Hz)
Human Walking	2–3
Car engine compartment	200
Refrigerator	240
HV AC vents in an office building	60
Kitchen blender casing	121
Cloth dryer	121
The instrument panel of the car	13

Source: [1].

6.1.1 CANTILEVER

6.1.1.1 Unimorph and Bimorph Structures

Generally, piezoelectric energy harvesting involves vibration sources with low amplitude accelerations or low frequencies. To harness energy from such mechanical sources, a thin and flat contour is favorable since it allows the piezoelectric specimen to react promptly to the host. Additionally, this type of configuration is advantageous in limiting the inclusive weight and dimensions of the device. As per statistics, cantilever geometry is one such design that is most used due to the simplicity of architecture. Also, the resonant frequency of its primary flexural modes is lower than the other vibration modes of the piezoelectric specimen. It is possible to build a thin layer of piezoelectric ceramic into a cantilever by fixing it on a non-piezo layer, mostly a conductor to work as an electrode. One of its ends is fixed to exploit the flexural mode of the structure and this architecture is named "unimorph cantilever" since one layer of active piezoelectric material is used (Figure 6.1(a)). However, a cantilever can also be fabricated by incorporating two slim layers of piezoelectric material onto the metallic layer to derive more power output per unit. This is termed a "bimorph cantilever' (Figure 6.1(b)), since it involves two active layers. It is convenient to improve the power delivered in the bimorph structure by connecting the layers either in series or in parallel (series connection increases the output voltage while parallel connection enhances the output current). In 2001, a group of researchers proposed energy harvesting from foot movement and incorporated it in different locations of the shoe (Figure 6.2(a)). In addition, Figure 6.2(b) displays the voltage and power outputs of the piezoelectric constructions on optimized resistive loads. The device was tested under normal walking speed, that is, 0.9 Hz (approximately) as is evident from the spikes of the outputs. In this work, the authors have used a bimorph configuration instead of unimorph at the heel area, which along with efficient power conditioning circuitry delivered around 1.3 mW of continuous power at a walking frequency of 0.8 Hz [2].

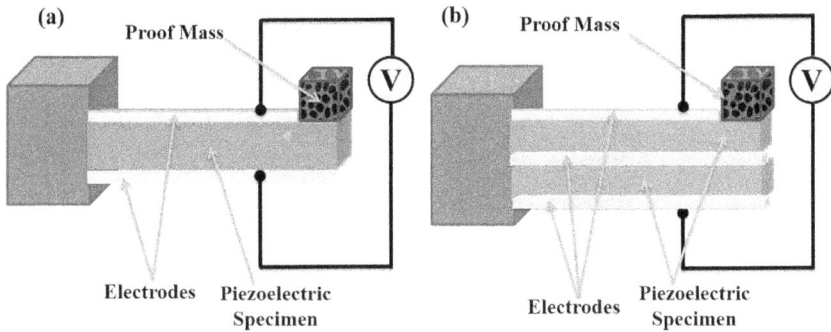

FIGURE 6.1 Schematic diagram showing construction of (a) unimorph and (b) bimorph cantilever piezoelectric generators.

6.1.1.2 Need for Proof Mass

It is well known that cantilever-type energy harvesters possess relatively simple architecture and relatively uncomplicated processing technology. In spite of that, cantilevers suffer from the limitation of a narrow range of resonant frequencies. The resonant frequency (f_r) of a simply assisted cantilever beam is given by the equation [3,4],

$$f_r = \frac{\upsilon_n^2}{2\pi} \frac{1}{L^2} \sqrt{\frac{EI}{mw}} \qquad (1)$$

Where, E = Young's modulus, I = moment of inertia, L = length, w = width, and m = mass/ length of the cantilever beam. $\upsilon_n = 1.875$ represents the eigenvalue for the fundamental vibration mode. Due to the limited resonant frequency range, it is often noticed that no resonance takes place if the frequency of vibration of the cantilever is marginally away from the ambient frequency. This in turn leads to less efficient power conversion. Two ways have been suggested in the past to resolve this issue: (a) adding a proof mass (Δm) or (b) extending the length of the beam. The former method is mostly preferred over the latter to lower the resonant frequency of the beam since an increase in the length might not fit into the idea of miniaturization of PEHs. For practical application purposes and modification of resonant frequency, it is more desirable to add a proof mass (Δm) to the free end of the cantilever. On inclusion of the proof mass (Δm), the previous equation is approximated as [4]:

$$f_r = \frac{\upsilon_n'^2}{2\pi} \frac{1}{L^2} \sqrt{\frac{K}{m_e + \Delta m}} \qquad (2)$$

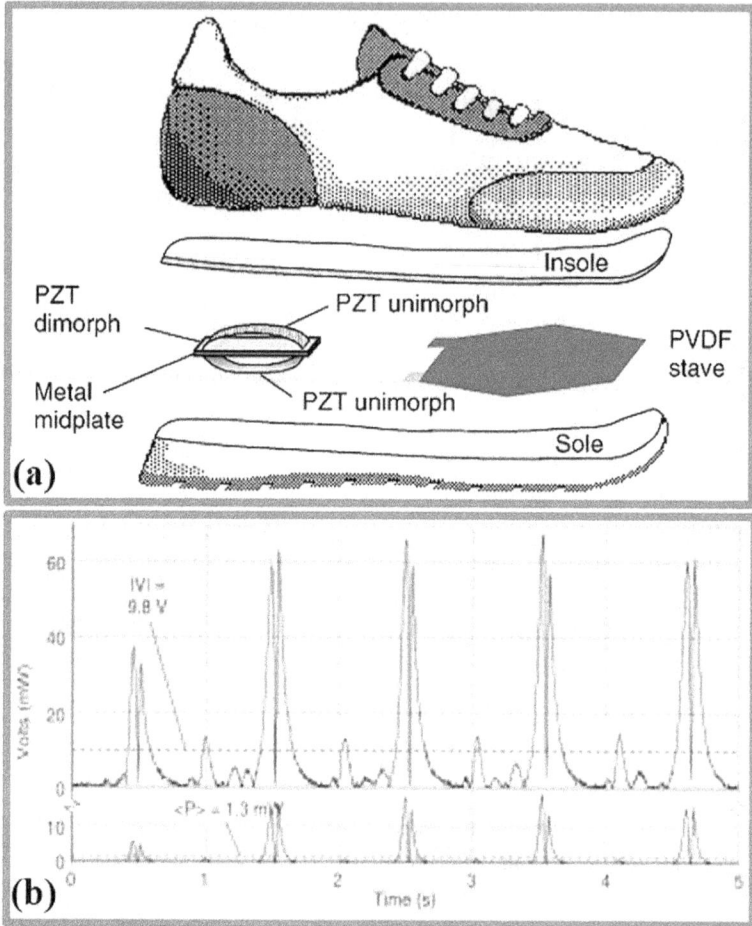

FIGURE 6.2   (a) Illustration showing two distinctive approaches for 31-mode piezoelectric energy harvesting from shoes – a PVDF stave below the ball of the foot and a PZT dimorph below the heel; (b) voltage and power waveforms under brisk walking conditions for optimum loading of PVDF stave.

Source: [2].

Here, $v_n'^2 = v_n^2 \sqrt{\dfrac{0.236}{3}}$, $m_e = 0.236\ mwL$ is the cantilever mass, Δm = proof mass, K = effective spring constant of the cantilever. In this regard, Roundy and his co-workers [5] ascertained the direct dependence of cantilever power efficiency on proof mass. This means, in order to improve the output power of the cantilever, Δm must be maximized within the acceptable range of design parameters inflicted by the beam strength and resonant frequency. There are many examples that depict an enhancement in the power efficiency and broadening of the resonant frequency range by different modifications in the proof mass. In 2008, Shen et. al. [6] devised a microelectromechanical system (MEMS) using PZT unimorph cantilever with Si proof mass. The device could deliver a power density 3272 μW/cm^3 and an optimal resistive load 6 kΩ (resonant frequency ~ 461.15 Hz) when the beam and integrated Si proof mass possessed a dimension 4.8 mm × 0.4 mm × 0.036 mm and 1.36 mm × 0.94 mm × 0.456 mm respectively. Besides, the accommodation of proof mass on the cantilever beam is at the expense of the active piezoelectric layer in conventional designs. To address this issue, a group of researchers from Toronto developed a cantilever type PEH having curved L-shaped proof mass that not only improved the power density but also lowered its fundamental frequency by 20–30 percent as compared to conventional block-shaped mass cantilevers. The L-type shape generated power (average) 350 μW and a power density of 1.45 mW/cm^3. This was almost 68 percent greater than the conventional type. Further, with a walking speed of 3 km/hour, the device could produce a power of 49 μW [7]. Recently, Moon et al. [8] proposed a new methodology for broadening the bandwidth of cantilever-type PEH by tuning the proof mass. In the first step, they explored the theoretical possibilities and then modeled the prototype of a broadband harvesting device. The fabricated device exhibited two voltage peaks at two different resonant frequencies slightly separated from each other (with different load resistances). The extracted power was considerably high, and the resonant bandwidth was much broader than any conventional cantilever PEH (Figure 6.3).

6.1.2 Cymbals and Diaphragms

Apart from cantilever structures, piezoelectric energy harvesters with circular architectures, including cymbal and diaphragm-type transducers, are also important. Cymbal-type harvesters are most essential in applications involving high-impact forces and were first introduced by Newnham and his group [9] in 1992. Typically, it consists of a ceramic disc made out of a piezoelectric material sandwiched between two metallic end caps that are bonded using any adhesive polymer such as epoxy (Figure 6.4). The end caps are usually made up of steel owing to its greater yield strength as compared to brass or aluminum. High values of yield strength are essential for greater load-bearing capability of the transducer [10]. Kim et al. [11] in their work described that steel caps can amplify the axial stress applied to the cymbal structure and convert it into radial stress in the piezoelectric PZT disc. Hence, the modes d_{33} and d_{31} contribute to the charge generation in transducer. For the cymbal architecture, the effective d_{33} is given by [12]:

$$d^{eff} = d_{33} + A\left|d_{31}\right|, \text{ here A is the amplification factor.} \qquad (3)$$

The amplification factor A can fall in the range of 10–100 depending on the design of the cymbal, i.e., dimensions of the cavity and end cap which directly affects

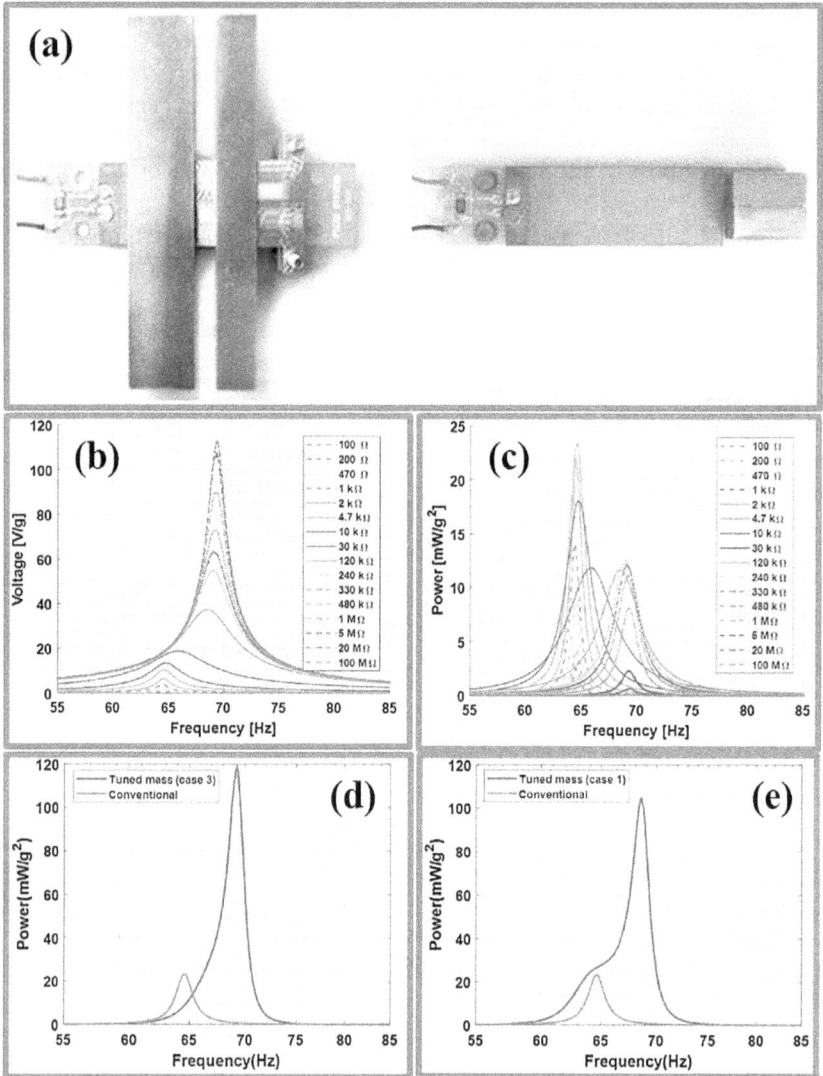

FIGURE 6.3 (a) Top view of the proposed piezoelectric energy harvester with tuned mass and conventional energy harvester; (b) voltage and (c) power response of the conventional harvester as a function of frequency for different load resistance; comparison of (d) maximum power and (e) bandwidth among the proposed and conventional device.

Source: [8].

the effective d_{33}. The piezoelectric cymbal possesses a very high transduction rate along with large displacements. A is usually described by the expression:

$$A = \frac{r_b(r_b - r_t)}{2t_h(t_p + 2t_m)}, \quad r_b, r_t, t_p, t_m \text{ are explained in Figure 6.4.} \tag{4}$$

Finally, the open-circuited voltage (V_{OP}) and the output power (P) derived from equations (3) and (4) are given by:

$$V_{OP} = \frac{Q}{C} = \frac{d^{eff}}{\varepsilon_{33}\varepsilon_0} \times \frac{Ft_p}{A} \tag{5};$$

here, Q is the charge and A is the area of the piezoelectric element.

$$P = \frac{1}{2}(d^{eff} \times g^{eff})\left(\frac{F}{A}\right)^2 \tag{6};$$

here, g^{eff} is the effective piezoelectric voltage constant and F = applied force.

The above equations give an idea that the output power delivered by a cymbal energy harvester depends directly on the transduction coefficient ($d \times g$)and applied force while inversely dependent on amplification factor A. Consequently, both the structure and nature of materials are two main factors affecting the power output of cymbal PEH. Based on the theoretically optimized mode of energy harvesting devices, deep insight into their working mechanism can be better understood in terms of electromechanical coupling, and the energy transfer process is highly essential. A critical analysis of improving the different parameters may lessen the working-inherent frequency gap to maximize the output power at f_r (resonant frequency) in a vibration surrounding. For instance, the frequency of the cymbal

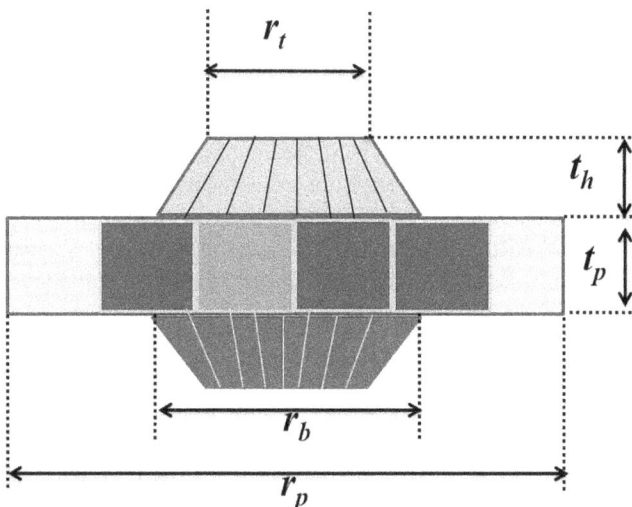

FIGURE 6.4 Schematic diagram of cymbal type piezoelectric energy harvester.

transducer is greatly influenced by the structure of the end caps or the bonding (epoxy) material. Even a minute asymmetry can result in a mismatch of resonant frequencies of both the metallic caps and even a double resonance peak arises in the frequency response. Besides, cymbal structures are known for their low-power applications, but their use in high-power ultrasonics is limited. This is because the epoxy layer degrades on driving the device at high power levels owing to high stress generation. Such an effect ultimately reduces the operating life of the device [13]. To address this issue, Bejarano et al. [14] made a comparative study of two cymbal transducer configurations for ultrasonic application. One of them was a conventional cymbal designed by Newnham [9] and the other consisted of piezoceramic bonded to a metallic ring. In order to maximize the mechanical coupling factor, they placed the end-caps with a large flange on the metal ring and attached them with screws. In addition to that, the end-caps were concave in shape to bear high pressures though it reduced the output displacement. This new design performance is better than the traditional ones with advantages of high axial shift for a small radial movement of the disc and constancy of end-cap movement (by increase in added mass). In another study, Zhao et al. [15] made a detailed study on the effect of different dimensions of cymbal structure on its efficiency and coupling coefficient for pavement energy harvesting. This was done using finite element analysis. The optimum geometry of cymbal for maximum efficiency was found to be: (i) total diameter ~32 mm; (ii) cavity base diameter ~22 mm; (iii) end-cap top diameter ~10 mm; (iv) cap steel thickness ~0.3 mm; (v) cavity height ~2 mm, and (vi) piezo-ceramic thickness ~2 mm. The output power and voltage were obtained to be 1.2 mW and 97.33 V at a vehicle load frequency of 20 Hz.

Further, it is observed that a combination of cymbal with other structures such as cantilever or multilayer stack prove to be useful for obtaining high efficiency of the piezoelectric energy harvesters under high load. Xu et al. [16] observed that the resultant structure on integrating cantilever with cymbal could overcome each other's limitations. Generally, cantilevers with a beam made up of piezoelectric material along with a proof mass work at low resonant frequency. However, non-uniform mechanical stress produced in the beam during bending vibrations hamper the conversion efficiency of the piezo-materials near its tip. On the other hand, cymbal PEH not only possesses high energy conversion efficiency but also high resonant frequency. This is unsuitable from the point of view of harvesting energy from the ambient environment. In this regard, they combined both the architectures to fabricate a composite energy harvesting system known as CANtilever Driving Low frequency piezoelectric Energy harvester (CANDLE). The CANDLE consisted of a cantilever and a couple of cymbal transducers is shown in Figure 6.5. With this unique configuration, the authors of this work could retain low resonant frequency along with high output power simultaneously. Experimental data indicate a decrease of leap point from 70 m/S^2 to 30 m/S^2 when the proof mass changed from 4.20 gm to 1.05gm.

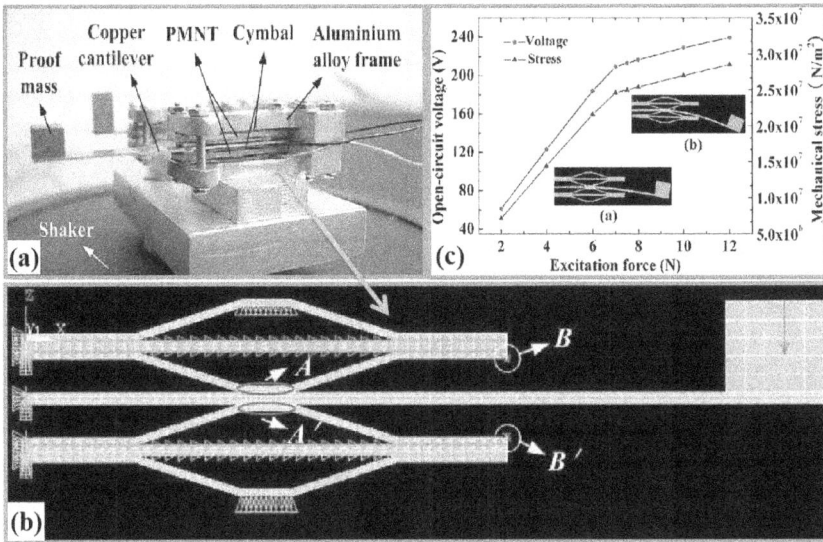

FIGURE 6.5 (a) Prototype device of piezoelectric CANDLE structure consisting of a cantilever beam with two cymbal transducers; (b) finite element model of CANDLE configuration; (c) dependence of open circuit voltage and mechanical stress on static force in the transducer structure (simulated).

Source: [16].

Piezoelectric circular diaphragm (Figure 6.6) is based on the same working principle as cantilever. It constitutes a piezoelectric ceramic in thin circular disc form bonded on a metal shim followed by clamping on both the edges. The major difference between circular diaphragm and cantilever is that the latter is clamped at one end only. Sometimes at the center of the diaphragm, a proof mass is attached to create pre-stress condition. This is because pre-stress within the piezo-ceramic is known to enhance the performance of the energy harvester at low-frequency. Another way to produce pre-stress in the ceramic material is to design a sandwich structure by pressing the circular diaphragm between two dissimilar metallic layers followed by heating and cooling to room temperature. Dissimilar thermal expansion coefficients lead to rolling of the whole structure, introducing pre-stress in the piezoelectric diaphragm. Besides, conventional diaphragms usually operate in 31 mode identical to piezoelectric cantilevers. However, in order to exploit the 33 mode of piezoelectric ceramic, NASA has developed a special type of interdigitated spiral electrode pattern. As shown in Figure 6.7, the positive and negative electrodes are alternatively patterned into the center of piezoelectric disc. This kind of diaphragms is called "Radial Field Diaphragm" (RFD) [17, 18]. RFDs are found to deliver 3–4 times higher output power than conventional diaphragms. In this regard, Shen et al. [19] reported PZT diaphragms with spiral interdigitated

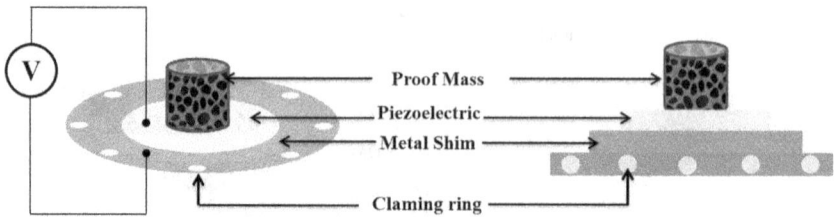

FIGURE 6.6 Schematic diagram showing the construction of a circular diaphragm piezoelectric energy harvester.

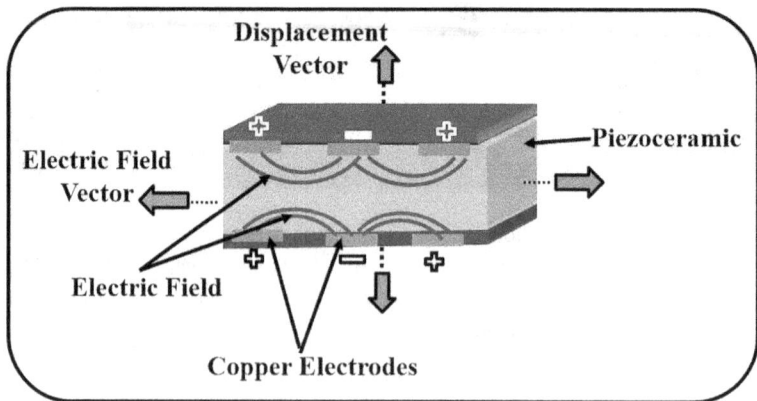

FIGURE 6.7 Schematic diagram showing interdigitated positive and negative spiral electrode pattern.

electrodes, which demonstrated the lowest resonant frequency of 1.56 Hz due to its small size. In addition to that, the power output obtained was in nanowatt under 1 g acceleration. Nevertheless, the power density was equivalent to that of 33-mode cantilevers and cymbal transducers. Furthermore, Anderson and his co-workers [20] described in their work a novel design of wide-band acoustic transducers, which utilize radially pressurized PVDF film having conducting electrodes on both upper and bottom surfaces. They compared this arrangement to a very common flexure-type transducer with a flat diaphragm (comprising of a piezoelectric layer, that is, PZT, and passive elastic layer). It was noticed that pressurized diaphragms (PVDF film) without the passive elastic layer demonstrated the same levels of displacement per unit electric field as compared to conventional flexural type transducer with PZT active layer. This was an interesting outcome since the piezoelectric coupling coefficient of PZT is approximately 100 times more than PVDF. In another recent report, Heke et al. [21] discussed about piezoelectric acoustic transducer constructed by encapsulating circular AlN and SiN diaphragms in air-filled cavity. The diaphragm was optimized for low minimum detectable pressure (MDP) using analytical finite element analysis. Finally, the experimental result for transducer of diameter 175 μm on a 400 × 500 × 500 μm³ die showed structural resonance at 133 kHz and 552 kHz in water and air respectively. The sensitivity at 10 Hz–50 kHz range was found to be 1.87 μV/Pa both air and water.

6.1.3 STACK

The third most important category of piezoelectric energy harvesters is stack PEH, constructed by stacking many piezoelectric films along the applied electric field (Figure 6.8). This is the device of choice in applications needing large energy output and high loads. However, the fabrication process of stacked structures is a bit complex. Initially, the ceramic powders, binder, dispersant, plasticizer and solvent are mixed together thoroughly to form a slurry. The slurry is then tape casted into thin films followed by drying and then screen printed with electrodes. Finally, stacking of these ceramic films are done in an ordered manner and subjected to sintering. The fabrication of stacked structures is shown in Figure 6.9. Stacked PEH can be connected either in parallel or in series with a motive of improving the power output as per the requirement for applications. Series connection of the piezoelectric stacks generates high open circuited voltage capable of igniting devices such as a lighter. On the other hand, parallel stacking leads to high current and power for electricity supply. When compared to a cantilever PEH with approximately the same size and weight, the power (electrical) and power density delivered by stack PEH in resonance as well as off-resonance modes is higher

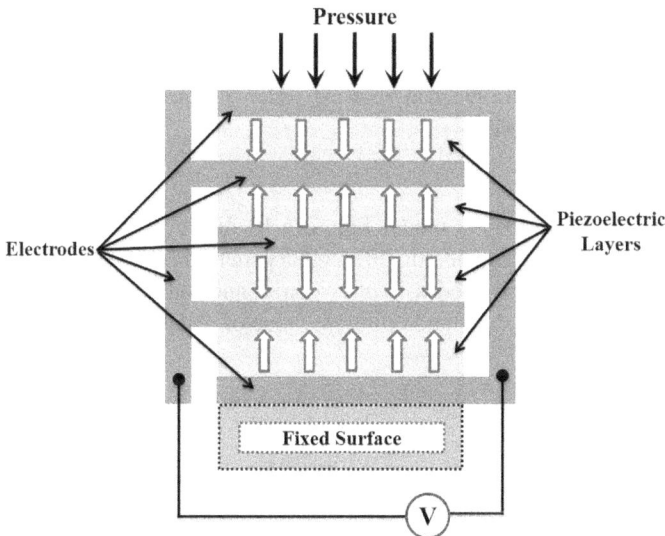

FIGURE 6.8   Schematic diagram showing stack piezoelectric energy harvester.

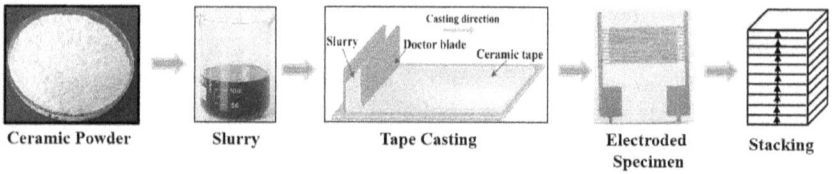

FIGURE 6.9 Schematic diagram showing different processes involved in the fabrication of stack piezoelectric energy harvester.

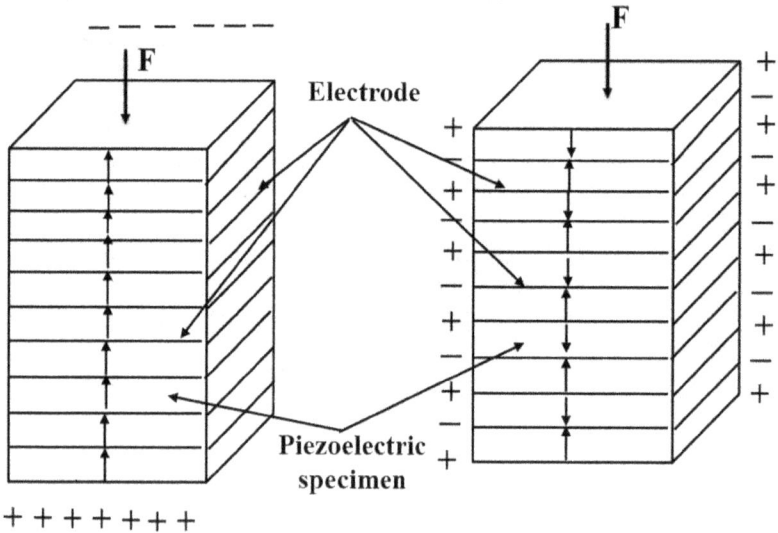

FIGURE 6.10 Schematic diagram showing series and parallel connections of stack piezoelectric energy harvesters.

because of excitation of d_{33} mode. Figure 6.10 displays the series and parallel connections of stack piezoelectric energy harvesters. If the distance between the positive and negative electrodes is nt_c when connected in series, the maximum value of output power (P_{MAX}) and optimum value of resistor (R_{OPT}) in series can be expressed as [12]:

$$P_{MAX} = \frac{d_{33}^2 F^2 \omega n t_c}{4\sqrt{2} A \varepsilon_{33}}$$ (7)

$$\text{and } R_{OPT} = \frac{n t_c}{\sqrt{2} \omega A \varepsilon_{33}}$$ (8)

Here, ω frequency of the force applied $[F = F_{max} \sin(\omega t)]$ vertically on the stack; t_c and A are the thickness and area of the piezoelectric specimen respectively; and $n=$ number of layers in the stack.

Parallel electroding of the stacked architecture results in a polarization direction opposite to the adjacent layers on application of a force in vertical direction. If the distance between both the electrodes is given by t_p, and the complete electrode area is given by nA, then P_{MAX} and R_{OPT} in parallel can be expressed as:

$$P_{MAX} = \frac{n^2 d_{33}^2 F^2 \omega n t_p}{4\sqrt{2} A \varepsilon_{33}} \tag{9}$$

$$R_{OPT} = \frac{t_p}{\sqrt{2} \omega n A \varepsilon_{33}} \tag{10}$$

As per equations (7) and (9), the maximum deliverable power in both cases of stacking, i.e., series and parallel are directly proportional to the specimen thickness t_p, d_{33}^2 and F^2 while inversely proportional to dielectric constant ε_{33} and area A of the specimen. So it is obvious to conclude that more thickness and less area would contribute to better piezoelectric output, but this idea is not suitable from the point of view of miniaturization. However, to obtain desirable output, proper choice of piezo- material with appropriate dimensions is necessary. Multilayer stacks are generally employed to generate high power levels. In this regard, Xu et al. [22] studied the response of PZT stack operated in 33 mode (electric field and/ or mechanical stress ᴨ polarization direction). They focused on the crucial energy harvesting properties such as generated electrical energy or power, mechanical energy conversion efficiency, deliverable power to the resistive load and energy transferred to the supercapacitor. The piezoelectric stack comprised of 300 layers of Navy Type II (CeramTec SP505) PZT plates of 0.1 mm thick alternating with pure silver internal electrodes of 0.1 µm thick. Passive layers (without electrodes) of 1 mm thick were also placed at the end. Capacitance of the piezoelectric stack was measured to be 2.5 µF at 1 kHz. The power and power density generated by this stack transducer were much greater than that of a weight and size cantilever type PEH. Recently, Khalili et al. [23] utilized stacked PEH for harvesting mechanical energy from roadways. With 500 kΩ external load resistance and 66 Hz load frequency, a maximum output voltage of 95 V to 1190 V was obtained from the stack. The harvested power (RMS power output = 9 to 1400 mW) was sufficient enough for powering a microprocessor. The merits and demerits of different structural configurations are given in Table 6.2 [24].

6.2 OPERATION MODES

The energy harvesting performance of piezoelectric materials are greatly dependent on their polar axis and the direction of applied force with respect to the polar axis. The polar axis is usually represented by '3' while the directions orthogonal to the

TABLE 6.2
Structural configurations with their merits and demerits

Configuration	Features/Merits	Demerits
Unimorph/bimorph cantilever	• Uncomplicated architecture • Cost effective fabrication • Low resonant frequency • Output power proportional to proof mass • Good mechanical quality factor	• Incapable of resisting high impact force
Circular diaphragm	• Suitable for pressure mode operation	• More stiff than cantilever of similar dimensions • High resonant frequencies
Cymbal transducer	• Greater energy output • Bears high impact force	• Applications restricted to high magnitude vibration sources
Stack PEH	• Withstands high mechanical load • Works well with pressure mode operation • Higher output from d_{33} mode	• High stiffness

polar axis are represented as '1' or '2.' When the stress is applied along the direction of polar axis, that is, '3', it leads to 33 mode and when it is applied perpendicular to the polar axis (mostly '1'), it leads to 31 mode. The 33–and 31-modes are the most conventional vibration modes utilized in case of energy harvesters as displayed in Figure 6.11. It is generally observed that the piezoelectric properties of the PEH in 33-mode are much higher than 31-mode resulting in better electromechanical coupling coefficient [25, 26]. Again, piezoelectric energy harvester in 33-mode also supports interdigitated electrode (IDE) that gives greater output voltage as compared to 31-mode owing to its low capacitance [27].

6.2.1   31-MODE

One of the most common modes in which samples can be conveniently fabricated in a wide range of shapes and sizes is the 31-mode. This mode is relatively simpler to fabricate since the PEH operating in this mode are associated with uncomplicated top and bottom electrodes (TBE). Apart from that, the microstructure of the piezo-layers is tuned according to the structure and type of the bottom electrode, resulting in improvement of deliverable power. Wang et al. [28] investigated on 31-mode energy harvester which basically consisted of a piezoelectric patch on a stainless steel or brass substrate adhered to each other using epoxy. After structural optimization, the voltage output and deliverable energy were calculated. They noticed that the output performance of the PEH was greatly influenced by

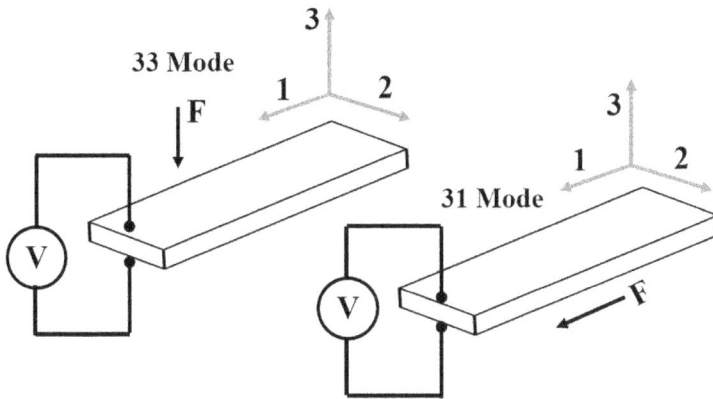

FIGURE 6.11 Schematic diagram showing conventional operation modes in case of piezoelectric energy harvesters.

the thickness as well as the modulus ratio between the piezo-layers and substrate. Further, a peak in output voltage was observed on a diverse thickness ratio, while the voltage increased initially and then attained saturation on changing the modulus ratio. Earlier, Singh et al. [29] examined the performance of piezoelectric ZnO based vibration energy harvester. The basic structure of the harvester consisted of a fixed free type cantilever with proof mass attached at the free end. ZnO piezoelectric layer was sandwiched between two metal electrodes and attached to the top of the cantilever beam. The authors of this work used FEM tool Coventor Ware to estimate the resonant frequency, optimum load resistance, stress and power generated. After optimizing the PEH, the resonant frequency was measured to be 235.38 Hz. In addition to that, the open circuited voltage output was recorded as 306 mV for 0.1 g harmonic acceleration. Similar structure was also developed another group of researchers using PZT piezoceramic. The PEH was fabricated to resonate at suitable frequencies from an external vibration source. The cantilever harvester was operated both in 33-mode and 31 mode and it was found that 33-mode generated 20 times higher output voltage as compared to 31-mode [25]. The following section describes the operation of PEH in 33-mode.

6.2.2 33-MODE

33-mode piezoelectric energy harvester involves rather complicated interdigitated electrodes (IDE) first reported by Wilkie et al. [30] and Hagood et al. [31]. In the literature, the reports on 33-mode of operation are comparatively less than 31-mode. Construction of IDE requires very fine patterning as compared to TBE, which needs simple rectangular shape patterning. Another limitation of device operating in 33-mode is the necessity of high poling voltage. This is because of the higher finger spacing in case of IDE that of the spacing between the electrodes in TBE. Such high voltages pose risk of device failure and electrical breakdown. Further, interdigitated electroding also affects the output power of the PEH, but it is typically less than 31-mode. Besides to discuss the structural design for conversion of vibration energy into electrical energy, let us consider the cantilever PEH as the displacement is easily generated in the perpendicular direction. As illustrated in Figure 6.11, on application of stress, electric charge is induced either perpendicular (31-mode) or parallel (33-mode) to it. As a result of accumulated charges, the open circuited voltage V_{OC} derived in different modes of operation is related to the applied stress σ_{ij}, voltage constant g_{ij} and distance between the electrodes G_e as:

$$V_{OC} = \sigma_{ij} g_{ij} G_e; \text{where } g_{ij} = \frac{d_{ij}}{\varepsilon_r \varepsilon_0} \tag{11}$$

Here d_{ij} is the piezoelectric constant; ε_r and ε_0 relative dielectric constant and absolute permittivity of vacuum. The most well-known piezoceramic PZT has demonstrated almost twice the higher piezoelectric constant in 33-mode as compared to 31-mode. Therefore, g_{ij} is also predicted to be twice in 33-mode than that of 31-mode. Again, from equation (11) it is quite clear that the open circuited voltage directly depends on the distance between the electrodes. In case of 31-mode, it is the spacing between top and bottom electrodes while for 33-mode, it is the spacing of electrode fingers. Considering the case of thin film PEH, G_e is much less for 31-mode than 33-mode resulting in higher output voltage in 33-mode. In contrast, 31-mode may be more useful in high current generation. In view of output power which is product of voltage and current, better metrics are shown by 31-mode of operation than 33-mode owing to low capacitance in the latter mode. In contrary, 33-mode PEH exhibit improved performance for actuator applications [32]. With all these considerations, Choi et al. [33] fabricated a thin film energy harvesting MEMS (micromechanical systems) sensors to monitor large social and environmental infrastructures in remote areas. It was designed in the form of a bimorph cantilever with interdigitated Pt/Ti electrodes. In 33-mode, this cantilever of area 170×260 µm generated 1 µW of continuous power to a 5.2 load resistance at 2.4 V DC. In another interesting project, piezoelectric free standing acoustic transducer in the form of circular diaphragm was operated in 33-mode. The diaphragm was associated with spherical electrode which allowed in-plane deformation to be converted into out-of-plane deformation to generate acoustic waves. The sensitivity of the diaphragm in such configuration (33-mode) was found to be 126.21 µV/Pa at 1 kHz, which is approximately 20 times higher than sandwich 31-mode transducer [34].

6.3 SUMMARY

This chapter provides extensive information on the working mechanism, device configurations and operational modes of piezoelectric energy harvesters. Though the PEH have undergone different stages of development from bulk to micro to nano, cantilevers still remain the most employed configurations owing to their simplicity of architecture and suitability in low-frequency environments for MEMS. Among the other configurations are cymbals, diaphragms and stacks. Diaphragms are almost similar to cantilevers except that they are fixed at both the ends and proof mass is attached in the center. Diaphragm PEH is mostly utilized in acoustic applications. Besides, cymbals play an important role in applications involving high impact force because of their high load bearing capacities. Nevertheless, stacked PEH offers flexibility in achieving desired levels of output power under high load by adjusting the number of piezoelectric layers and different electrode connection modes. Apart from the different configurations, vibration modes also play a crucial role in modulating the performance of the piezoelectric energy harvesters. Two of the most dominant vibrational operation modes are 31-mode and 33-mode. There are three principal factors contributing to power generation in both the modes: spacing between the electrodes, electrode area and piezoelectric voltage constant. It is found that 31-mode is most prevalent due to simplicity of electrode configuration and ease of fabrication. But the deliverable power is much higher in case of 33-mode in spite of complex interdigitated electroding. Finally, the content of this chapter will definitely guide the researchers to plan better design strategies, taking in to account the different rubrics in order to develop a proper PEH.

REFERENCES

1. Varadha, E. and Rajakumar, S., 2018. Performance improvement of piezoelectric materials in energy harvesting in recent days-a review. *Journal of Vibroengineering*, 20(7) pp. 2632–2650.
2. Shenck, N.S. and Paradiso, J.A., 2001. Energy scavenging with shoe-mounted piezoelectrics. *IEEE Micro*, 21(3), pp. 30–42.
3. Li, X., Shih, W.Y., Aksay, I.A. and Shih, W.H., 1999. Electromechanical behavior of PZT-brass unimorphs. *Journal of the American Ceramic Society*, 82(7), pp. 1733–1740.
4. Yi, J.W., Shih, W.Y. and Shih, W.H., 2002. Effect of length, width, and mode on the mass detection sensitivity of piezoelectric unimorph cantilevers. *Journal of Applied Physics*, 91(3), pp. 1680–1686.
5. Roundy, S. and Wright, P.K., 2004. A piezoelectric vibration based generator for wireless electronics. *Smart Materials and Structures*, 13(5), p. 1131.
6. Shen, D., Park, J.H., Ajitsaria, J., Choe, S.Y., Wikle, H.C. and Kim, D.J., 2008. The design, fabrication and evaluation of a MEMS PZT cantilever with an integrated Si proof mass for vibration energy harvesting. *Journal of Micromechanics and Microengineering*, 18(5), p. 055017.
7. Li, W.G., He, S. and Yu, S., 2009. Improving power density of a cantilever piezoelectric power harvester through a curved L-shaped proof mass. *IEEE Transactions on Industrial Electronics*, 57(3), pp. 868–876.
8. Moon, K., Choe, J., Kim, H., Ahn, D. and Jeong, J., 2018. A method of broadening the bandwidth by tuning the proof mass in a piezoelectric energy harvesting cantilever. *Sensors and Actuators A: Physical*, 276, pp. 17–25.
9. Sugawara, Y., Onitsuka, K., Yoshikawa, S., Xu, Q., Newnham, R.E. and Uchino, K., 1992. Metal–ceramic composite actuators. *Journal of the American Ceramic Society*, 75(4), pp. 996–998.

10. Kim, H.W., Batra, A., Priya, S., Uchino, K., Markley, D., Newnham, R.E. and Hofmann, H.F., 2004. Energy harvesting using a piezoelectric "cymbal" transducer in dynamic environment. *Japanese Journal of Applied Physics*, *43*(9R), p. 6178.

11. Kim, H.W., Priya, S., Uchino, K. and Newnham, R.E., 2005. Piezoelectric energy harvesting under high pre-stressed cyclic vibrations. *Journal of Electroceramics*, *15*(1), pp. 27–34.

12. Li, L., Xu, J., Liu, J. and Gao, F., 2018. Recent progress on piezoelectric energy harvesting: structures and materials. *Advanced Composites and Hybrid Materials*, *1*(3), pp. 478–505.

13. Ochoa, P., Pons, J.L., Villegas, M. and Fernandez, J.F., 2006. Advantages and limitations of cymbals for sensor and actuator applications. *Sensors and Actuators A: Physical*, *132*(1), pp. 63–69.

14. Bejarano, F., Feeney, A. and Lucas, M., 2014. A cymbal transducer for power ultrasonics applications. *Sensors and Actuators A: Physical*, *210*, pp. 182–189.

15. Zhao, H., Yu, J. and Ling, J., 2010. Finite element analysis of Cymbal piezoelectric transducers for harvesting energy from asphalt pavement. *Journal of the Ceramic Society of Japan*, *118*(1382), pp. 909–915.

16. Xu, C., Ren, B., Liang, Z., Chen, J., Zhang, H., Yue, Q., Xu, Q., Zhao, X. and Luo, H., 2012. Nonlinear output properties of cantilever driving low frequency piezoelectric energy harvester. *Applied Physics Letters*, *101*(22), p. 223503.

17. Bryant, R.G., Effinger Iv, R.T., Aranda Jr, I., Copeland Jr, B.M., Covington Iii, E.W. and Hogge, J.M., 2004. Radial field piezoelectric diaphragms. *Journal of Intelligent Material Systems and Structures*, *15*(7), pp. 527–538.

18. Li, H., Tian, C. and Deng, Z.D., 2014. Energy harvesting from low frequency applications using piezoelectric materials. *Applied Physics Reviews*, *1*(4), p. 041301

19. Z. Shen, S. Liu, J. Miao, L. S. Woh, and Z. Wang. Proof mass effects on spiral electrode d_{33} mode piezoelectric diaphragm-based energy harvester. 2013. IEEE 26th International Conference on Micro Electro Mechanical Systems (MEMS), pp. 821–824. IEEE, 2013.

20. Wixom, A.S., Anderson, M.J., Bahr, D.F. and Morris, D.J., 2012. A new acoustic transducer with a pressure-deformed piezoelectric diaphragm. *Sensors and Actuators A: Physical*, *179*, pp. 204–210.

21. Hake, A.E., Zhao, C., Ping, L. and Grosh, K., 2020. Ultraminiature AlN diaphragm acoustic transducer. *Applied Physics Letters*, *117*(14), p. 143504.

22. Xu, T.B., Siochi, E.J., Kang, J.H., Zuo, L., Zhou, W., Tang, X. and Jiang, X., 2013. Energy harvesting using a PZT ceramic multilayer stack. *Smart Materials and Structures*, *22*(6), p. 065015.

23. Khalili, M., Biten, A.B., Vishwakarma, G., Ahmed, S. and Papagiannakis, A.T., 2019. Electro-mechanical characterization of a piezoelectric energy harvester. *Applied Energy*, *253*, p. 113585.

24. Mishra, S., Unnikrishnan, L., Nayak, S.K. and Mohanty, S., 2019. Advances in piezoelectric polymer composites for energy harvesting applications: a systematic review. *Macromolecular Materials and Engineering*, *304*(1), 1800463.

25. Jeon, Y.B., Sood, R., Jeong, J.H. and Kim, S.G., 2005. MEMS power generator with transverse mode thin film PZT. *Sensors and Actuators A: Physical*, *122*(1), pp. 16–22.

26. Park, J.C., Lee, D.H., Park, J.Y., Chang, Y.S. and Lee, Y.P., 2009, June. High performance piezoelectric MEMS energy harvester based on d_{33} mode of PZT thin film on buffer-layer with $PbTiO_3$ inter-layer. In *TRANSDUCERS 2009-2009*

International Solid-State Sensors, Actuators and Microsystems Conference (pp. 517–520). IEEE.

27. Park, J.C., Park, J.Y. and Lee, Y.P., 2010. Modeling and characterization of piezoelectric d_{33}–mode MEMS energy harvester. *Journal of Microelectromechanical Systems*, *19*(5), pp. 1215–1222.

28. Wang, Q., Dai, W., Li, S., Oh, J.A.S. and Wu, T., 2020. Modelling and analysis of a piezoelectric unimorph cantilever for energy harvesting application. *Materials Technology*, *35*(9–10), pp. 675–681.

29. Singh, R., Pant, B.D. and Jain, A., 2020. Simulations, fabrication, and characterization of d_{31} mode piezoelectric vibration energy harvester. *Microsystem Technologies*, *26*(5), pp. 1499–1505.

30. Wilkie, W.K., Bryant, R.G., High, J.W., Fox, R.L., Hellbaum, R.F., Jalink Jr, A., Little, B.D. and Mirick, P.H., 2000, June. Low-cost piezocomposite actuator for structural control applications. In *Smart Structures and Materials 2000: Industrial and Commercial Applications of Smart Structures Technologies* (Vol. 3991, pp. 323–334). SPIE.

31. Hagood, N.W., Kindel, R., Ghandi, K. and Gaudenzi, P., 1993, September. Improving transverse actuation of piezoceramics using interdigitated surface electrodes. In *Smart Structures and Materials 1993: Smart Structures and Intelligent Systems* (Vol. 1917, pp. 341–352). SPIE.

32. Kim, S.B., Park, H., Kim, S.H., Wikle, H.C., Park, J.H. and Kim, D.J., 2012. Comparison of MEMS PZT cantilevers based on d_{31} and d_{33} modes for vibration energy harvesting. *Journal of Microelectromechanical Systems*, *22*(1), pp. 26–33.

33. Choi, W.J., Jeon, Y., Jeong, J.H., Sood, R. and Kim, S.G., 2006. Energy harvesting MEMS device based on thin film piezoelectric cantilevers. *Journal of Electroceramics*, *17*(2), pp. 543–548.

34. Shen, Z., Lu, J., Tan, C.W., Miao, J. and Wang, Z., 2013. d_{33} mode piezoelectric diaphragm based acoustic transducer with high sensitivity. *Sensors and Actuators A: Physical*, *189*, pp. 93–99.

7 Applications
Sources and Devices

7.1 ENERGY HARVESTING FROM FLUIDS

The flow of fluids in our surroundings (air flow/liquid flow) carries lot of motion energy, which offers a huge scope for PEH. Despite the fact that the use of this energy for the production of grid power has been well established for a long time, it is still less explored for wireless sensor networks and electronic devices. In recent years, small wind turbines with rotor diameters as low as a few meters have been created. These systems are best suited to supply off-grid power locally in areas that are inaccessible to grid power, including residences and workplaces in distant locations, boats and trailers, telemetry equipment, and roadside signage. However, even the smallest commercially available turbine (rotor diameter ~1m; power ratings ~100 W) is quite large by the standards of energy harvesting for miniature electronic devices and sensor networks. This section intends to assess the state of art of energy harvesting miniature devices that convert the flow energy of water and air into useful electrical energy. In the literature, this type of energy conversion has been described employing a variety of forms, including cantilevers, windmills, plates, films, membranes, flags, and discs, among others. We discuss a few of them in the following sections.

7.1.1 WINDMILL TYPE HARVESTER TO CAPTURE AIRFLOW

Miniature windmill structures are a good inception for harvesting of airflow energy, and several research groups have worked toward it. This form of energy is especially important due to its wide availability and the wind's enduring nature. However, the matter of concern is the relatively low wind speeds close to the ground due to boundary layer effects and trees like corporeal hindrances. The first demonstrations on small-scale windmills were made in the early twenty-first century. Federspiel and Chen in 2003 [1] designed an AC generator having a simple rotor fan of diameter 10.2 cm combined with a DC motor. The output of the generator was rectified using a three-phase bridge. This whole structure expected to deliver DC power of 8 and 28 mW at an air speed of 2.5 m/s and 5.1 m/s respectively. Followed by this work,

DOI: 10.1201/9781003317289-7

Shashank Priya in 2005 [2] reported the performance of theoretically modeled piezoelectric bimorphs at lower frequencies than the piezoelectric resonance. This model was evaluated in three steps: (i) based on the bending beam theory for bimorph, the open circuit voltage response of the transducer under AC stress was calculated; (ii) the equivalent circuit of a capacitor connected to a resistive load used the computed open circuit voltage as input; and (iii) finally the model outputs were validated by the measured response of a prototype windmill. The framework of the windmill (piezoelectric) was almost analogous to a traditional windmill with active blades made up of 10 piezoelectric bimorphs arranged along the circumference. The bimorphs possessed a dimension of $60 \times 20 \times 0.6$ mm^3 and a free length of 53 mm. Capacitance and resonant frequency of the bimorph were estimated to be 170 nF and 65 Hz respectively. The bimorph transducers were so arranged that they oscillate between the stoppers due to the influence of wind flow. The oscillations were guided by the cam shaft gear mechanism, and such motions can continuously produce electricity. This prototype was tested, and a power of 7.5 mW could be derived at a wind speed of 10 mph across a resistive load of 6.7 kΩ. Again in 2007, Priya and his group [3] presented two different designs with an improved version of a small-scale windmill. The first model consisted of blades in the form of cup vanes (depth = 1.2 inch; diameter = 2.3 inch; arm length = 3.5 inch). The cup vanes were made up of 12 bimorphs epoxied along the circumference by removable clamps. Further, under the influence of wind flow, the vanes rotated a shaft having three triangular bumps. These bumps directly hit the bimorphs at every $10°$ making them vibrate and generate voltage. However, it was associated with many disadvantages, such as discomfort in packing bimorphs due to their cylindrical shape, the significant decrease in lifetime of bimorphs due to continuous impact of the bumps and, most importantly, the momentum was not high enough to allow up to four bimorphs in the structure with wind speed below <7 miles/h,. In the second model, they improvised the windmill with pre-stressed blades (three in number). This addressed all the disadvantages of the former model. The use of multiple vanes could increase surface area, capture more wind energy, and generate more power. The different advantages of the latter prototype included reduction in force due to two sets of bimorphs placed opposite each other, optimization of frequency by adjusting the gear ratio, large surface area of vanes for capturing more wind energy, and high rotational inertia to prevent windmill clogging. Similar work was carried forward by several other research groups [4–6]. Besides, a unique piezoelectric windmill was presented by Kan et al. [7] to harvest wind energy at low and wide ranges of speeds (Figure 7.1). The windmill consisted of piezo-cantilever transducers (piezoelectric membrane fixed on top of a substrate), which were excited by rotating magnets, fan blades, and tabulated framework. These piezoelectric cantilevers were attached to the outer side of tabulate framework along with an exciting magnet fixed to their free end. The repulsive interaction between the exciting and excited magnets drove the cantilevers by minimizing the impact of exciting magnets on the structure. Furthermore, it was shown that there were various optimum rotary and wind speeds for the peak amplitude ratio and

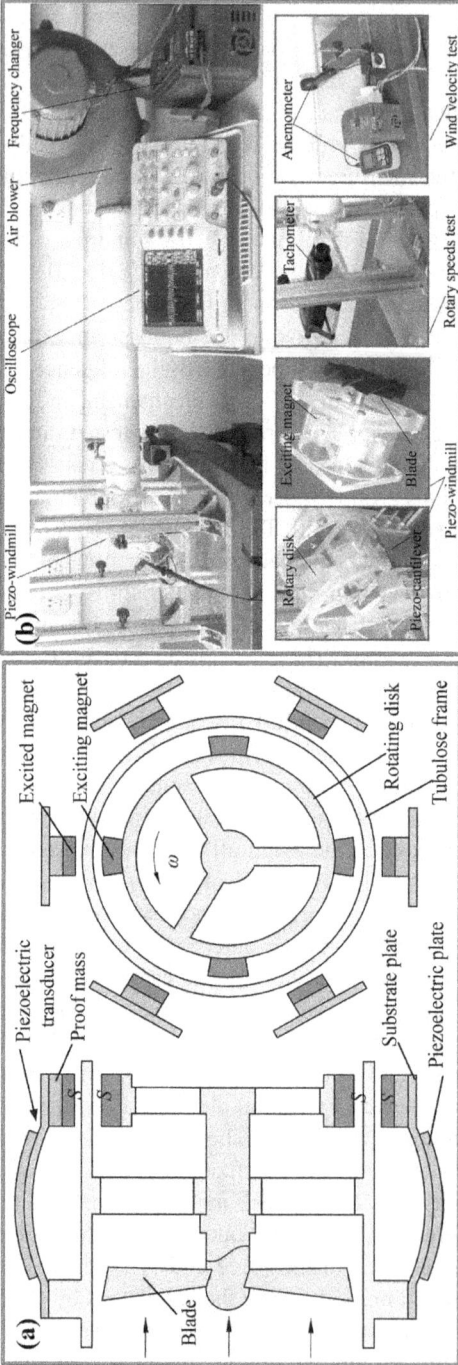

FIGURE 7.1 (a) Schematic diagram showing the structure and working of a piezoelectric windmill; (b) Photograph displaying a lab scale piezo windmill and various tests conducted.

Source: [7].

generated voltage. As the number of exciting magnets rises, the number of ideal rotary/wind speeds falls. In order to obtain maximum, a reasonable number of exciting magnets generated the highest amplitude ratio/generated voltage.

7.1.2 FLUTTER TYPE HARVESTER TO CAPTURE AIRFLOW

Flutter may be described as the aeroelastic instability of a compliant immersed in a fluid flow. Flutter-type systems using piezoelectric technology are the second most important configuration for wind energy harvesting. Fluttering beams with attached airfoils, wake galloping (a fluttering beam in the wake of a fixed bluff), and galloping bodies are a few of the common constructions in this category (fluttering beams with attached bluff bodies). These harvesters are useful in overcoming some of the shortcomings of windmills (i.e., high manufacturing and maintenance costs, unnecessary complexity, a tiny scale's adverse scalability owing to viscous drag and friction). Besides, flutter mode is based on the aerodynamic effect given by the classic Karman vortex shedding principle [8, 9]. According to this principle, a lifting surface starts vibrating when it reaches a certain speed while in motion with the airflow. It oscillates and flaps continuously due to the involvement of inertial forces, elastic forces, and unsteady aerodynamics. Alternatively, the structure vibrates continuously under the influence of small disturbances. However, it stops flapping on reducing the wind speed below the critical speed (flutter wind velocity). Energy harvesting based on air-borne flutter mode is being researched by many groups with a hope of utilizing renewable sources in the future. In one of the works, Li and Lipson [10] designed a piezo-leaf (vertical stalk) generator that was able to convert energy from wind to electrical energy. The transducer used the principle of airflow-induced flapping motion and delivered improved output power as compared to conventional vibration generators. In 2007, Tan and Panda [11] tried to expose a piezoelectric beam to transverse airflow, which was mounted at an angle of 10 degrees. At the free end, a compliant plastic flapper was attached to improve the coupling wind-flow and beam. With this structure, a prototype harvester with dimensions $76.7 \times 12.7 \times 2.2$ mm^3 demonstrated an output power of 155 μW at a wind speed of 6.7 m/s (optimum speed). A nearly identical structure was developed and optimized with better output in 2011 [12]. An innovative design of novel leaf like PEH based on venation growth algorithm was developed by Wang and his group in 2018 [13]. This algorithm was based on a dicotyledonous plant's leaf prototype that has a netted distribution of veins (Figure 7.2). The authors of this work fabricated polyvinylidene (PVDF) leaf that was driven by vortex-triggered vibrations due to wind flow behind the bluff body. To probe into the energy harvesting performance of the leaf and its different components, wind tunnel experiments were conducted. The results obtained were quite interesting – the power output of the veined structures was 4 to 6 times more than the no-vein ones. The highest value of root mean square open circuited voltage was obtained to be 1.094 V under a wind speed of 11 m/s for leaf thickness of 110 μm. Recently, stalk leaf flutter type wind energy harvesters were developed by Hu and his group

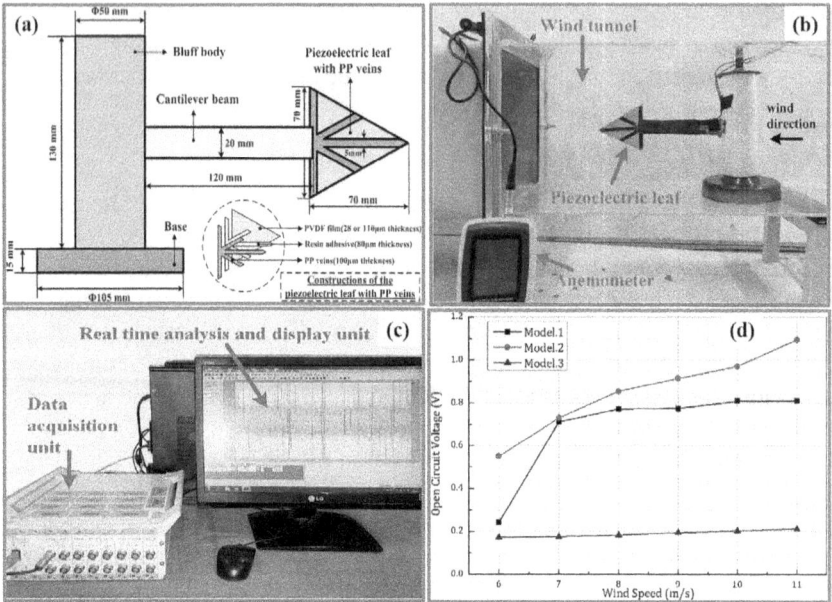

FIGURE 7.2 (a) Schematic diagram showing leaf-like structure of wind energy harvester; (b) experimental set up showing wind tunnel experimental system; (c) real time analysis and data acquisition unit; and (d) open circuit voltage obtained for three different models as described in reference.

Source: [13].

[14] in order to increase the efficiency of power and decrease the cut in speed. They used the frictionless hinge in the finite element model of the stalk and leaf system to implement the low-speed flutter mechanism. The prototype of the stainless steel stalk leaf inclined at angles 0°, 15°, 30°, 45°, 60°, 75° and 90° were fabricated and examined in wind tunnel. Further, the performance was estimated by fixing patches of macro-fiber composites on the root of the stalk. It was observed that an inclination of 30° demonstrated the lowest flutter speed. On the other hand, output power and voltage increased gradually with increase in air flow speed. Besides, 90° stalk leaf showcased abrupt velocity-power and voltage curves, having the maximum wind speed. This model exhibited a possible way to balance the efficiency with a reduction in speed.

7.1.3 HARVESTING LIQUID FLOW

Energy harvesting from liquid flow is among the least-researched areas in the current scenario as compared to the air flow, owing to the limited availability of source. However, liquid in the form of water holds great potential for providing ceaseless energy along with high-energy density contrast to wind-flow. Moreover, hydroelectric systems based on water are already popular all round the globe for generating electrical power. Recently, remote water sources are also gaining attention for small-scale energy harvesting to generate electric power. Installing piezoelectric systems in pipelines is one of the major leaps in harvesting energy from water flow. Many studies have been carried out in this regard by various researchers. Laser et al. [15] utilized a flow energy harvesting device consisting of a piezoelectric transducer in the form of a cantilever made of one or several layers of piezoelectric material. The cantilever was placed in a converging–diverging channel in which its vibrations were guided by the fluid flow. This system delivered an energy of 20 mW at 20 L/min flow rate and 165 kPa pressure drop. In another work, a bicylinder vortex-induced vibrator (VIV) was designed by a group of workers to harness energy from the flow of water. Load resistance, diameter of the cylinder, and speed of water flow were identified to be the most crucial factors in optimizing the power (electrical) output. The optimum harvested power was determined to be 21.86 μW [16]. Taking this work forward, the same group considered an upright cylinder for harvesting energy that undergoes VIV due to water flow. A maximum of 84.49 μW output power was extracted from this system [17]. Moreover, a tandem type system comprising of two similar PEHs was investigated by Shan et al. [18]. Each of the harvesters was made using a circular cylinder along with a piezoelectric beam. Output power of the downstream harvester velocity dependent and was found to be 533 μW at a flow velocity of 0.412 m/s.

Sea waves or ocean waves carry a lot of sheer power that could easily exceed 5 kW/m of wave front. Harvesting this huge mechanical energy from sea waves acts as an effective alternative for a self-sustainable power source amidst areas where other power sources are unavailable [19]. In this regard, a new device named as an energy-harvesting "Eel" using piezoelectric polymers or their composites, have gained much importance. Eel generators utilize the regular trail of travelling vortices at the back of bluff body to strain piezoelectric specimens, which leads to undulating motion. Such kinds of wobble motion mimic the movement of a natural electric eel. Since the commercially available piezo-polymers exhibit attractive properties at low cost, eels constructed out of them are also cost-effective and are easily scalable. These devices can also generate power in the range of mW to W based on their size and the flow velocity of the surroundings. Taylor et al. [20] in 2001 developed a typical eel structure using commercial PVDF (a piezoelectric polymer) in β-phase. The interplay between the hydrodynamics of water flow and the mechanical energy conversion by the eel's structural components were thoroughly tested. Figure 7.3 represents a schematic diagram of the forces affecting the eel motion. The associated bluff body regularly sheded alternating vortices on either side of it with a specific frequency in a non-turbulent flow, which is mostly determined by the flow speed and bluff body width. The pressure difference created by the vortices causes the eel to move in an oscillating fashion. Finally, the piezoelectric polymer experiences a strain that causes low frequency AC voltage

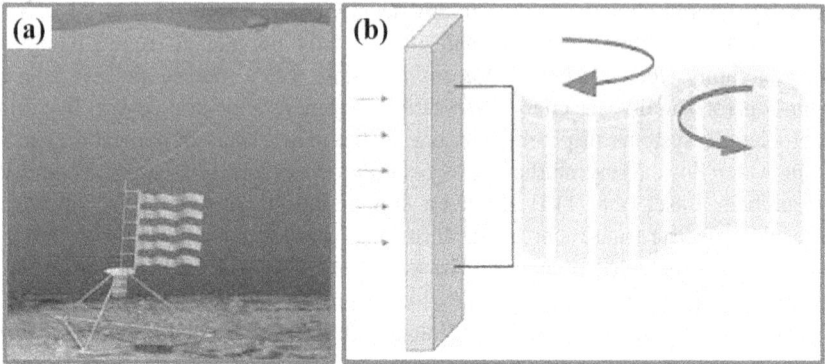

FIGURE 7.3 (a) Deployed eel structure under the water; (b) movement of the eel behind the bluff body.

Source: [20].

to be created along the electrode section. A central non-active core and a layer of active material on either side of the central layer are the only layers that are typically present in an eel body, though there may be more layers present. The core layer is often thicker and built from a flexible polymer with high bending moment to support the movement of other layers. The electrical power "P" generated by an undulating eel under defined flow conditions is given by,

$$P = \frac{\eta_1 \eta_2 \eta_3 A \rho V^3}{2};$$

here, η_1 represents the hydrodynamic efficiency that depends on the frequency matching of the oscillating eel with vortex-shedding frequency at the back of bluff body; η_2 is the efficiency of conversion of strain energy in to electrical energy by the piezoelectric polymer; η_3 is the extraction efficiency of electrical energy and is dependent on electrical losses in the resonant circuit; A = area of cross-section of the eel; ρ = density of water, and V = velocity of the water flow. Based on these principles, Taylor and his co-workers constructed a multi-eel system with 5 eels (dimensions: 52"× 6"× 400 μm). Similarly, Techet et al. [21] employed multiple eels stacked vertically behind a single bluff body to power long-endurance missions using sensors. Apart from studying the behavior of the eel stack, the effect of the length of a single eel was also investigated. It was found that when the length of the eel was longer, the radius of curvature near the head was less. This might have been caused by the added weight and overall resistance on the membrane. Reduction in curvature led to remarkably less strain-energy density and maximum available strain energy. Hence, eel designs with shorter material lengths and optimized electrode spacing might be more useful.

7.2   HUMAN BODY

Increasing use of portable electronic gadgets, including cell phones, global positioning systems (GPS), smart phones, portable digital assistants, laptops and so forth in our day-to-day life is leading to a massive increase in power needs. Usually the power demand of such devices is served by that need to be charged or replaced at regular intervals. This is a serious limitation and is detrimental for the lifespan of electronic devices. A safer alternative is to exploit the heat and mechanical motions of the human body to generate power. In particular, the kinetic energy from limb motion, muscle movement, pumping of heart, stretching of skin and so forth are some of the sources that present a unique opportunity to employ piezoelectric energy harvesting. The very initial effort in this direction was given by T. Starner in 1996, when he hypothesized a wearable computer powered by the energy from human body [22]. Since then, many researchers are working in this direction to develop new technologies for harnessing this energy with reference to on-body, or wearable, applications. On the other hand, piezo-technology-based human body energy harvesting can also benefit implantable active devices, including cardiac pacemakers, cardiac monitors, cardioverter defibrillators, and neurological brain stimulators. It can reduce the subsequent maintenance costs and reduce risks of failure. Hence, in this section, we will discuss the piezoelectric energy harvesting applications in implantable and wearable devices.

7.2.1   WEARABLE

Biomechanical motions of human body parts, like walking, jogging, finger movement; typewriting and elbow bending and so forth generate enough waste energy to power small-scale electronics. Among all biomechanical movements, motion of the lower limb (leg swing) can deliver maximum power, owing to higher torques as compared to others [23]. But extracting the waste energy from human body motion is relatively complicated and challenging. This is due to the ultra low frequency and multidimensionality of biomechanical movements. Piezoelectric technology has proved to be one of the effective methods of energy harvesting in such cases. A few of the examples are smart footwear, smart textile, smart skin and so forth. One of the earlier works by a research group at Massachusetts Institute of Technology (MIT) in this regard dates back to 1998. In this work, they incorporated PVDF stave and PZT unimorph into a standard jogging sneaker and accumulated sufficient energy across several steps. To judge the system performance, a small battery was laminated directly to the sole (as is done in case of usual LED flashing sneakers) and the lifetime of the shoe was compared to it. The results depict that approximately 10 nW of power was generated by the shoe with piezoelectric energy harvesting system, which lasted for approximately 2 years. The power generated is equivalent to 150 cm^3 of $LiCl_2$ batteries, that provide highest energy density among all Li-based cells [24]. Several advancements were made following this work to overcome the drawbacks of harnessing biomechanical movements. A very popular problem is related to the addition of metabolic cost due to a change in the locomotion pattern of the body part. Xia et al. [25] addressed the issue by creating a pair of energy-harvesting angle-sliding shoes without raising the metabolic costs. Similarly, Cha et al. [26] transformed a pair of slippers by removing the heel counter and tried to extract energy from them through bending/twisting. An interesting work was performed by Katsumura et al. [27] in which they equipped smart shoes with piezoelectric vibration energy harvesters. These energy harvesters were capable of generating operating wireless modules and in sole sensors through impulsive forces while running and working. Such technology is particularly important in sports shoes. Following this work, this research group has also fabricated prototype sports shoes containing resin housing with two piezoelectric energy harvesters, controller, power supply circuit, Bluetooth LE module, and others.

Flexible, bendable, stretchable, wearable, and breathable piezoelectric energy harvesters are in high demand for self-powered devices and micro-sensors in the area of health care services and human-computer interactive robotics. To achieve flexibility, the substrate should be flexible depending on the requirements of the electronic system. For instance, electronic skins are typically fabricated from flexible silicon-organic resin (silicone). On the other hand, flexible electronic displays are mostly made from polyethylene terephthalate (PET) due to lower requirement of flexibility. Recently, Fu et al. [28] prepared a stretchable electronic skin truly resembling tactile sensing and the mechanical property of human skin. Composites made from carbon black and silicone with high piezo resistivity was employed as the stimuli response substrate, which offered a product that was highly stretchable, super soft, had large-strain sensitivity and long-term reliability. In order to simulate the true mechanical behavior of human skin, highly conducting carbon fiber is used as an electrode and was designed (quasi-sinusoidal shape) so as to offer the excellent stretch qualities of e-skin (Figure 7.4). Also, the derived stress–strain curve of the device resembled a typical J-shape, almost equivalent to that of human skin. Besides, the fabricated electronic skin was capable of detecting the location and magnitude of dynamic stress distribution. In another article, Mokhtari et al. [29] fabricated knitted fibers using hybrid piezofibers of barium titanate nanoparticles and PVDF in 1:10 mass ratio. These knitted fibers in the form of a wearable energy harvester, could deliver an optimum output of 4 V and power density of 87 μW/cm^3. This value is almost 45 times higher than some previously reported figures. Interestingly, this harvester could charge a 10 μF capacitor in 20 seconds. Based on this wearable device, a knee sleeve prototype was designed and applied for monitoring real time precise health care (Figure 7.5). Further, the described processing technique was found to be scalable for synthesis of industrial-scale smart textiles.

7.2.2 IMPLANTABLE

Heart pacemakers, cardioverter defibrillators, heart monitors, brain and nerve simulators, and other implantable electronic devices provide ongoing diagnostic and therapeutic functions for a variety of disorders. These devices primarily rely

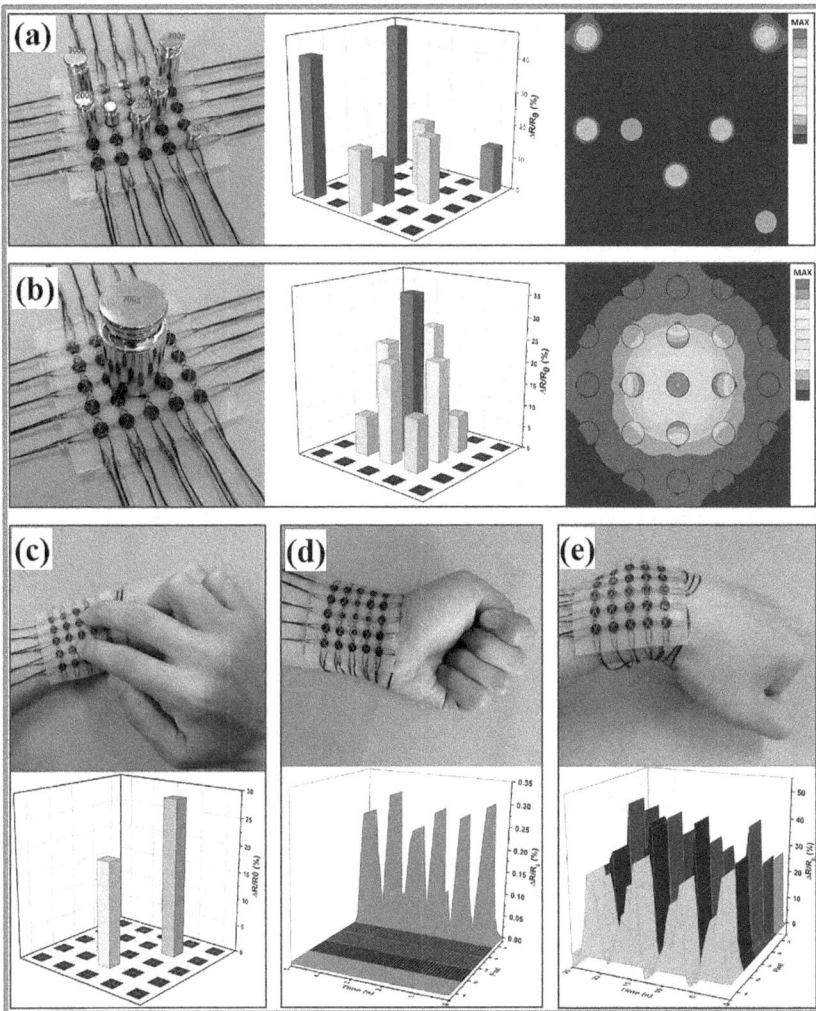

FIGURE 7.4 Relative changes in resistance (RCR) of the energy harvester monitored by pixels of the e-skin and finite element analysis under different loads (a) 100, 200 and 300 gm; (b) large 700 gm at the center; pixel records of RCR changes for the e-skin (c) touched by two fingers; (d) wrist artery blood pulse on closing the fist; (e) wrist bending.

Source: [28].

on battery technology, making surgical operations necessary to replace the spent batteries. A course of action like this has a substantial risk of morbidity and even mortality. In order to address this serious issue, low-power consuming implantable devices with a possibility of using the body's parasitic energy are prospective.

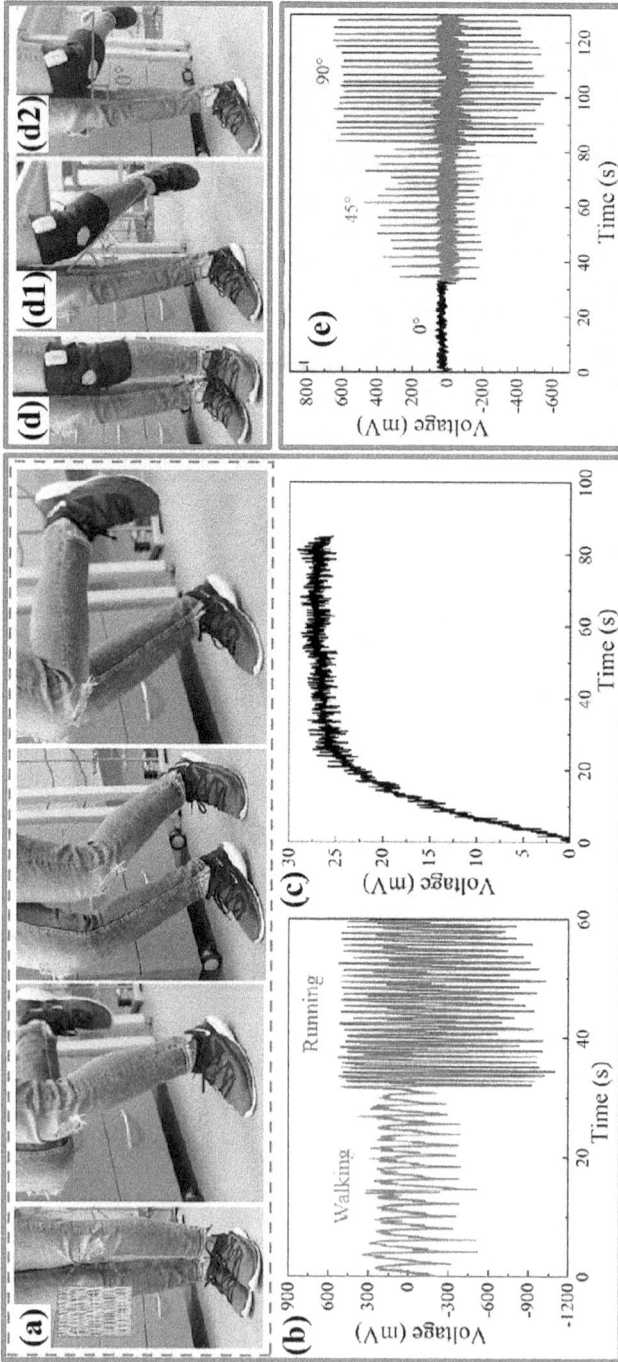

FIGURE 7.5 (a) Picture showing portable and wearable piezoelectric made from woven PVDF/BaTiO3 piezoelectric fibers and different bending positions; (b) output voltage from joint bending during walking as well as running; (c) voltage versus time graph during the charging of 10 μF capacitor on following 25 steps; picture of portable knee sleeve designed using woven PVDF/BaTiO3 piezo-sensor (d, d1 and d2) and its corresponding voltage–time graph (e) at a knee bending angle of (e) 0°, 45° and 90°.

Source: [29].

For example, the latest cardiac pacemakers could deliver 8 µW. Additionally, the human heart uses systolic and diastolic motion to propel blood circulation in order to provide steady, unending power throughout its lifespan. According to estimates, the typical human cardiac output is 1.4 W (product of aortic pressure and cardiac output). It suggests that a fresh and sustainable source of power for implantable devices can be generated using the energy of a beating heart. In this regard, Roger and his team [30] experimented with several animal models to capture energy from the motion of the heart and other organs. However, there are several risks, including pericardial tamponade, hemorrhage, impaired heart function brought on by forceful contractions, and cardiac vulnerability. However, barring all these disadvantages, many research groups have tried to employ the mechanical to electrical transduction feature of piezo-devices as a viable route for energy harvesting. Potkay and Brooks in 2008 [31] fabricated a structure consisting of a silicon sheet encapsulated using a small PVDF material enclosing a thin latex tube. The thin latex tube resembled an artery and was pressed manually to simulate arteriopalmus. A power of 16 nW and 1.2 V of output voltage were produced by the device. The PVDF film is suited for elastic deformation due to its good piezoelectric characteristics and low young's modulus. A flexible and implantable piezoelectric generator that harnesses the pulsing energy of the ascending aorta was tested for its viability and effectiveness in 2015 by Zhang et al. [32]. The piezoelectric generator was fabricated using an aluminum coated flexible PVDF film and was connected to two copper electrodes (Figure 7.6). A hollow latex tube with an inner diameter of about 12 mm and an outside diameter of about 17 mm was used to pack the entire device. Again, heart pumping was mimicked using an intra-aortic balloon pump. In order to simulate blood flow, it was put inside the latex tube and filled with saline. The in vitro study estimates an optimum voltage output (i.e., 10.3 V), current output (i.e., 400 nA) and power (i.e., 681 nW). The greatest voltage and current measured when the prototype device was wrapped around the pig ascending aorta were 1.5 V and 300 nA, respectively, with a heart rate of 120 bpm and blood pressure of 160/105 mmHg. A 700 ms instantaneous output power with a 77.8 percent duty ratio was recorded. Finally, the implanted generator was capable of charging an 1µF capacitor to 1 V in 40 seconds. In another appealing work, minimally invasive bioelectrical interfaces (i.e. soft implantable neurostimulators with programmable electrical-stimulation functionality) was explored by Chen et al. [33] in 2021. They initiated the electrical stimulation of peripheral nerves with soft PVDF-based thin film nanogenerator using ultrasonic pulses (Figure 7.7). Successful neurostimulation was achieved with sciatic nerves of rats by employing thin film generators of thickness 30 µm. Nevertheless, the stimuli control was methodically studied as a function of ultrasound parameters such as acoustic pressure, pulse width and pulse interval. Such a device presents a unique strategy of building a programmable battery-less neurostimulator that can provide real-time response to programmable external sources of energy.

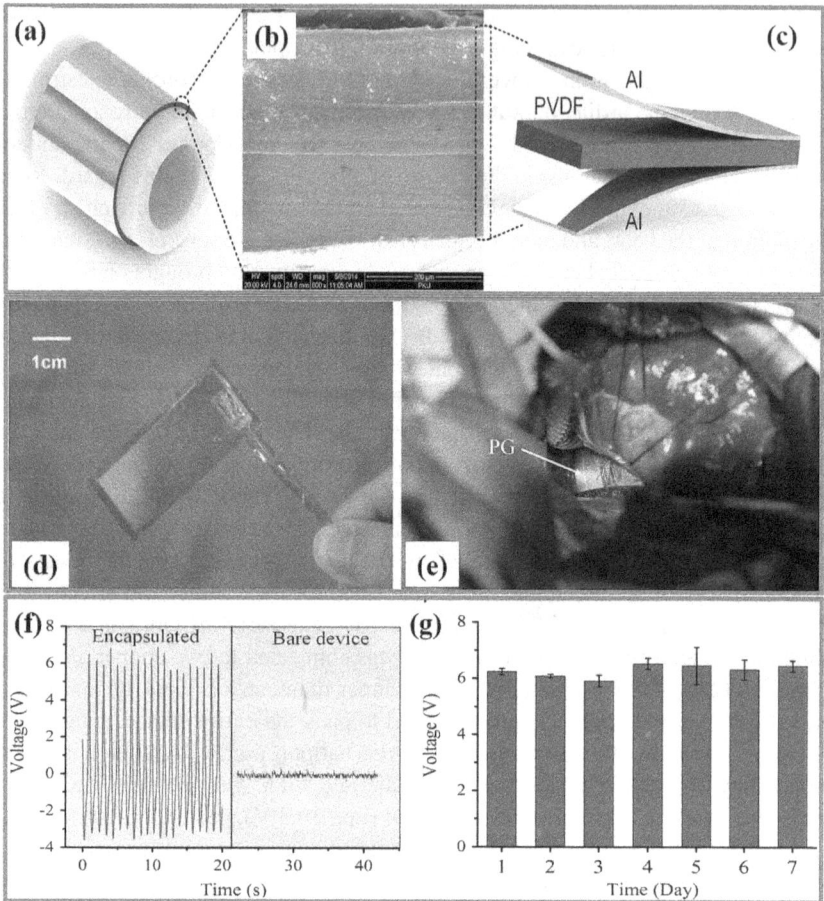

FIGURE 7.6   (a) Schematic figure showing the wrapping of piezoelectric generator over a latex tube; (b) SEM micrograph of generator; (c) a snapshot of the piezoelectric generator; (d) photograph of the generator sealed using PI tape; (e) wrapped piezoelectric generator implanted around ascending aorta; (f) output voltage waveforms of in vitro tests conducted for encapsulated and bare device in normal saline; (g) output voltage derived from the encapsulated device immersed in normal saline for one week.

Source: [32].

7.3   ANIMALS

Bio-logging systems have been installed on animals for a long time to measure some crucial factors related to the animal or its environment. Although these devices have been used since the 1930s, digital sensors, microcontrollers, and wireless data communication are becoming more and more common nowadays. Though these contemporary systems ensure better control over an assemblage of

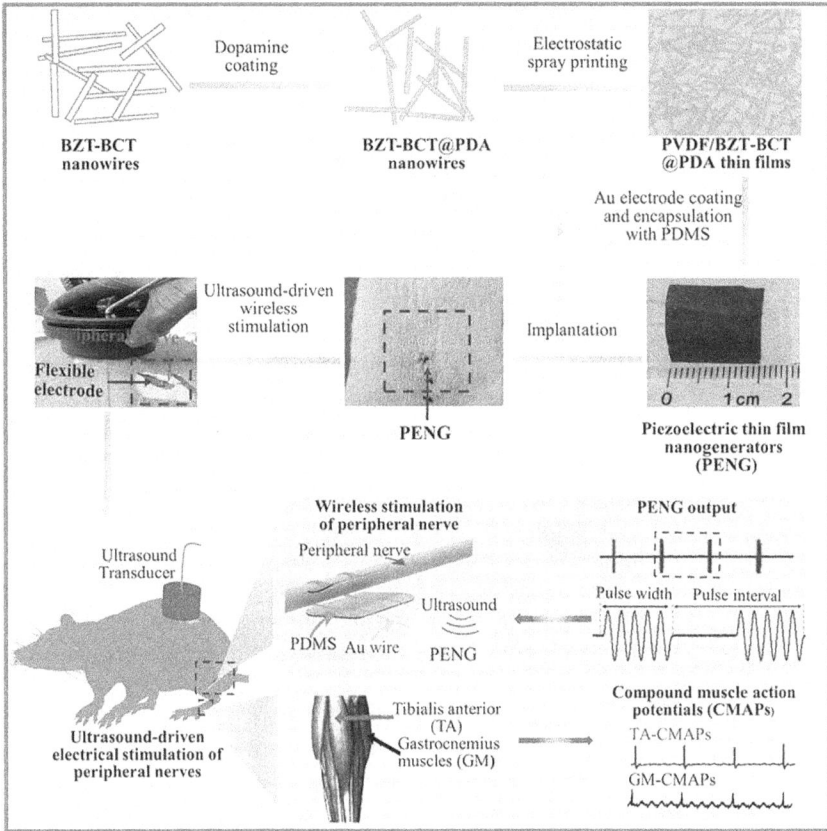

FIGURE 7.7 Detailed diagram showing the fabrication of PVDF/BZT-BCT@PDA piezoelectric thin film nanogenerator and wireless electrical stimulation of peripheral nerves in rats using these nanogeneratotrs remote driven by programmable ultrasound pulses.

Source: [33].

large data sets, power consumption remains one the greatest limitations. To resolve this issue, commercially available solar cells are employed to satisfy daily energy needs. However, it is not very advantageous for nocturnal, subterranean, or aquatic animals, or for animals that spend most of their time beneath dense forest canopies. Researchers investigating the habitat and activities of marine species require some self sustainable power source such as piezoelectric nanogeneraters, owing to the least availability of solar energy under water. Marine animals hover over wide swaths of oceans and seas, hence making data collection and tracking complex. Shafer and Morgan [34] stated that harnessing the fluid flow energy or variation of pressure as the animal dives could be two main energy sources, which can be harnessed easily. Piezoelectric technology could be one of the best methods of exploiting these phenomena. Recently, Quian et al.'s [35] conceptual design for a bi-stable piezoelectric energy harvester as a marine fish monitoring telemetry tag was influenced by biological systems. This self-powered tag might be attached to a fish's dorsal fin to track the habitats, populations, and aquatic environment of the species. The Venus Fly Trap, which uses a predatory shape transition to scavenge energy from fish movement and fluid flow, served as an inspiration for the design of this device. The initial underwater data indicate that the energy harvester could efficiently convert fish swings into electrical energy. The maximum deliverable energy obtained from this device was 17.25 mJ over 130 sec under 30° peak-to-peak swing and 1.5 Hz frequency using experiments of capacitor charging (Figure 7.8). On a related issue, Dr. E. Garcia and his Cornell University research team have begun a thorough investigation into the viability of generating power from aviation sources. According to preliminary research on flying insects, the flapping movement of the wings may generate roughly 40 mW of muscle energy, and between 1/2 and 1/3 of the insect's mass can be put onto it without having an impact on the stability of flapping. In 2008 [36], Reissman and Garcia made the discovery that it would be possible to capture the vibration energy of moth wings while they are in flight. They conceptualized the surgical incorporation of small generator and storage components along with neurological control unit into the pupa phase of the moth during its development. Following this, the same research group suggested that about 59 μW of power was produced by 0.292 gm of harvester mass on hawkmoth in untethered flight condition. This energy was enough to glow 196 mW LED every two seconds for duration of 29 μsec [37]. In 2015, Shafer et al. [38], proposed a method of harvesting power from the flight motion of birds and bats by mounting the energy harvester on their wings. Similar research works were performed by many other research groups, and this will definitely compel the use of piezoelectric harvesters to power the self-sustainable bio-logging tags and sensors.

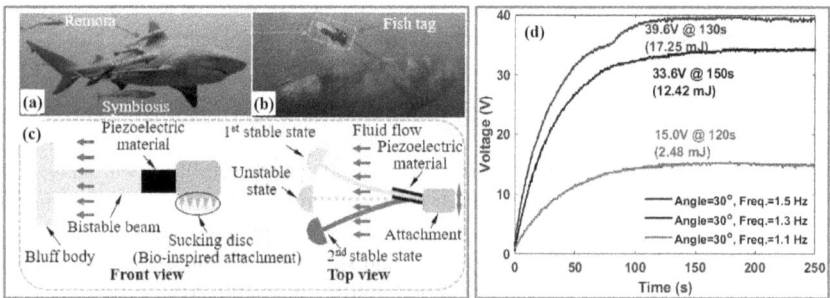

FIGURE 7.8 Design of a bio-inspired self powered fish telemetry tag: (a) symbiotic relation among remoras and shark; (b) externally deployed electronic tag on the dorsal fin of the shark; (c) concept design of the extension along with proposed bistable piezo-energy harvester with bluff body for improvement in structure-fluid interactions; (d) voltage versus time graph of charging a 22 μF capacitor by the harvester.

Source: [35].

7.4   INFRASTRUCTURE

Currently, energy management in smart cities is becoming more complex with growing energy demands of contemporary society. To resolve this problem, energy harvesting from manmade infrastructure and buildings may be advantageous. Artificial structures such as bridges, roads, and houses/tall buildings can be promising sources for vibrational/piezoelectric energy harvesting. The harvested energy can be used to power nearby wireless sensor systems, traffic lights, monitoring of vehicle speed, structural health monitoring and so forth. Commercial piezoelectric energy harvester tiles were installed in a key hub building at Macquarie University in Sydney, Australia, in an intriguing study by Li and Strezov, published in 2014 [39]. They created a model that recommended installing piezoelectric tiles with the greatest pedestrian mobility in 3.1 percent of the total floor surface. The total annual energy-harvesting capability of proposed pavement was estimated to be 1.1 MW h/year as per the simulation results, and it was sufficient enough to serve 0.5 percent of the total energy needed for the building. Subsequently, similar work was performed by Hwang et al. [40] in 2015, with the objective of designing piezoelectric tiles for harvesting energy from footsteps. It had three set of plates: an upper plate, a middle plate with piezoelectric modules, and a bottom plate with anchor points for the springs. Four cantilever beams with tip masses attached to the upper plate make up the piezoelectric modules. The whole structure is displayed in Figure 7.9. The derived power was optimized with the help of impedance matching technique and was measured to be 770 µW and 55 mW of RMS and peak power respectively. This value was much higher than that harvested from a shoe [41]. Further, an encouraging way of utilizing piezoelectric energy for the generation of electric power from high rise buildings under the influence of earthquakes and dynamic wind was proposed by Xie et al. [42] in the year 2013. The proposed harvester was based on recovering lost energy from a tuned mass damper's oscillations. Using piezoelectric patches, the tuned mass is supported across a vertical cantilever. However, the entire design was optimized, taking into account a number of factors, including the length and placement of the piezoelectric patch, its mass and radius, and the thickness ratio of the piezoelectric patch to the cantilever. Theoretically, maximum power conversion efficiency was found to be 28 percent. Later in 2015, the same research group was able to fabricate a prototype of the proposed harvester and place it on the roof of tall buildings. Under specific high-frequency seismic-caused building motion, the improved structure may generate 432.21 MW of RMS power [43]. In locations with heavy vehicle traffic, there is a great potential for energy harvesting from sources other than buildings, such as pavement deformation brought on by passing automobiles. In this regard, Jiang et al. [44] investigated upon a novel compression based PEH to harness power from vehicle traffic. The design had three multilayer stacks of piezo-material arranged in circular manner. Further, the laboratory scale prototype was tested using a shake table and maximum deliverable 2000 W/h power was obtained when 2,000 vehicles, each having a speed of 100 km/h, pass per hour. Similar works were also performed by many other research groups [45, 46]. Further, bridges also can

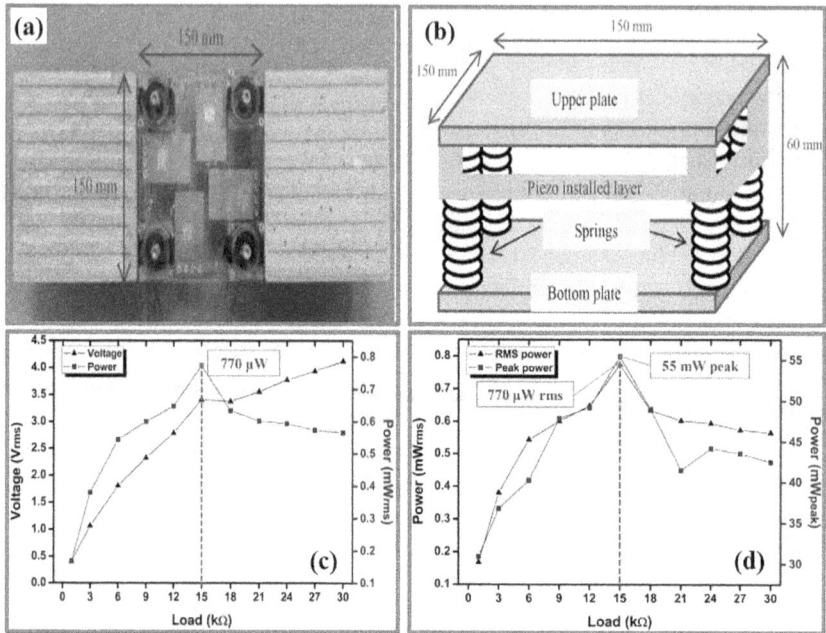

FIGURE 7.9 (a) Conceptual design and arrangement of real tiles and piezoelectric tiles; (b) structural design of piezoelectric tile; variation of (c) derived voltage and power and (d) RMS power and peak power from the piezo-tiles as a function of load.

Source: [40].

largely contribute to vibration energy harvesting. In the beginning, a case study was taken of a bridge in France, and its vibration characteristics were studied using several accelerometers. On the basis of these tests, the vibration frequency was determined to be very low, containing small amplitudes. According to experimental findings, 0.03 mW of electricity came from the harvester during a brief, random pulse train of vehicles crossing the bridge. The energy harvesting performance of piezoelectric bimorphs set to the natural frequency of the bridges and coupled with a vehicle–bridge system was recently evaluated by a group of researchers [47]. The sentence itself suggests that the researchers have conducted numerous experiments to enhance the efficiency with which the devices capture energy.

7.5 VEHICLES

Dissipation of energy in different parts of the vehicle, particularly, the suspension spring system, leads to reduction in the fuel efficiency of vehicles. As per reports, only 10–16 of fuel energy is utilized to power the vehicle and rest is lost in dissipation against road friction and air drag. P. Mucka [48] conducted a study in 2016 to precisely simulate the energy harvesting potentiality of suspension dampers in a typical passenger car. Since they used a comprehensive car model of an actual car, a tire-enveloping contact model and, most important, a big database of real road profiles, the author said that the results are extremely close to situations encountered in real life. As per his findings, the average power dissipated in the case of four dampers, attached to the wheels of a car moving at 60 km/h was in the range 24.6–25.9 W for AC and rigid pavements. However, the average waste power was calculated to be 24.6–57 W for the vehicle moving at 90 km/h. This waste energy, if harvested, could be useful in powering the onboard sensors and reducing the load on the vehicle's main battery, resulting in an increase in the efficiency of the vehicle. In 2008, Khameneifar and Arzanpour worked on energy harvesting from the deflection of car tires [49]. An average passenger car's usable tire deflection energy was calculated to be between 1040 and 1100 W. There were 14 commercial piezoelectric resonators ($9.20 \times 4.38 \times 0.99$ cm^3) installed within the tires, producing a combined output of 42 mW for the first resonant mode operation under standard conditions. Researchers also looked into the viability of using an inertial vibrating energy harvester to power a sensor module for usage with tires. Two mechanical stoppers and a bimorph cantilever built of PZT-ZNN layers made up the piezo-harvester (introduced limited mechanical strain to generate power from radial oscillations of the tire). Broadband operation, small volumes and sizes were some of the crucial parameters while designing the harvester. Finally, the improved design generated 31 W of power across a 330 k resistive load at 80 Hz and 0.4 g RMS base excitation [50]. Recently, Maurya et al. [51] showed that it is possible to harvest strain energy from car tires to power wireless data transmission with higher frame rates and self-powered strain detection. The testing was done in a laboratory environment using an organic polymer based piezoelectric patch at several frequencies (referring to different vehicles speeds). The results suggest a peak power of 580 µW at 16 Hz. This power was used for wireless data transfer, and it was discovered that the output power was adequate to directly illuminate 78 LEDs (Figure 7.10). Finally, the field testing showed that under the same load, there was no noticeable change in voltage output (from the sensor) due to change in topography. However, under typical loads and speeds, a distinct change in output voltage was seen, which was consistent with the simulation results (Figure 7.11).

7.6 MULTIFUNCTIONAL

Conventional energy harvesting systems using piezoelectric technology are designed to generate electrical energy from mechanical motions to power small electronic devices without offering any added function. Such devices might be thought of as ad hoc in nature or as add-on elements to the host design, which could result in undesired mass loading effects and unnecessary space waste. However, the idea behind a multifunctional energy harvesting modality is that the structure used to collect the energy should be able to offer some extra benefits at the same time, such as storing the scavenged energy or supporting mechanical load. Such

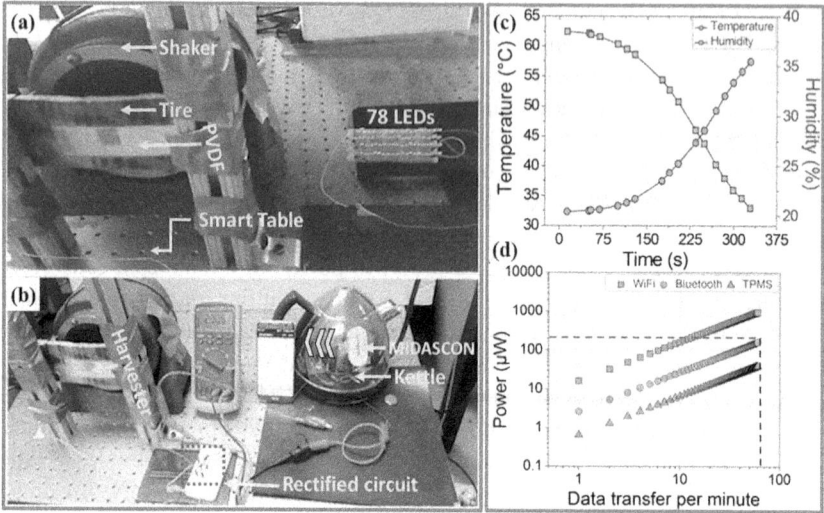

FIGURE 7.10 (a) Piezoelectric patch illuminating 78 LEDs; (b) demonstration of transfer of wireless data via stored power from the piezoelectric patch mounted on the tire section; (c) transfer of test temperature and humidity data using the stored energy; (d) power management analysis for different technologies such as Wifi, Bluetooth and TPMS.

Source: [51].

types of technological advancement has originated over a past few years and is still in the nascent stage. It was only in 2010 that Anton et al. [52] presented a multifunctional strategy to pair a thin film battery with a piezoelectric harvester for energy storage. This special technique was able to collect strain energy and store it in battery layers at the same time. The same authors' group expanded on this idea by creating multipurpose energy harvesting systems of unmanned aerial vehicles (UAVs) by integrating them into the wing spar [53]. The blueprint of the device was so made that energy was initially harvested due to the vibration of the wings during flight. Thereafter, the energy is stored for later use. Wang and Inman enhanced this concept further in 2013 [54] by including actuation capabilities with the aim of providing gust relief as well as energy gathering. In order to disable the gust forces on the wing, self-powered gust alleviation used vibration energy from the kinetic energy of the wings to activate a piezoelectric device. The vibration amplitude of the first and second modes could be reduced by 28 dB and 37 dB, respectively, according to theoretical simulation results. Apart from that, combination of energy harvesting and vibration attenuation using locally resonant metastructures is a matter of recent study taken up by researchers. Particularly for essential applications like low-frequency vibration attenuation in flexible constructions, locally resonant metastructures are composed of metamaterials components that enable bandgap creation at wavelengths significantly higher than lattice size. Sugino and Erturk [55] attempted to combine the fields of energy

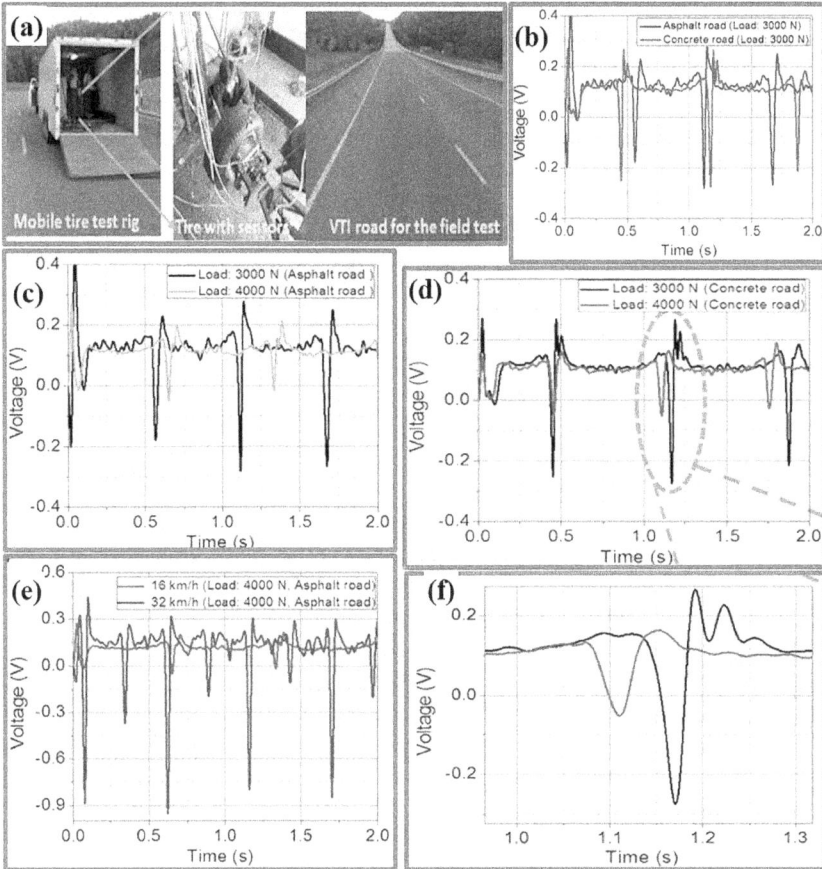

FIGURE 7.11   (a) Experimental set up for field tests; (b)-(d) generated waveforms due to deformation of the tyre at a speed of 16 km/hr on different terrain and loads at a tire pressure of 193 kPa; (e) comparative waveform change at 16 km/hr and 32 km/hr; (f) enlarged image of single cycle of waveform with different wheel loads.

Source: [51].

harvesting and locally resonant metamaterials in order to build multifunctional devices that showed the ability to both generate low-power voltage and attenuate vibrations. Despite of the promising results, developing a fully integrated locally resonant metastructure with intricate energy harvesting circuitry that performs at its best is still a problem that needs to be solved.

7.7   MULTISOURCES

Traditionally, only one specific type of energy source is intended for each energy harvesting technology. For instance, thermoelectrics can collect thermal energy, piezoelectrics can harvest kinetic energy, and photovoltaic energy harvesters can only harvest light. However, the output power of single-source energy harvesters frequently falls short of meeting the power needs of wireless sensor networks, in part because the source is unstable or has intermittent availability. Therefore, one approach to solve this problem is to develop multisource energy harvesting systems, which may simultaneously collect energy from a variety of ambient energy sources and strengthen the system as a whole. During the last few years, the scientific community has been working on multisource energy harvesters that can combine piezoelectric vibration harvesting with other types of harvesting sources. Magoteaux et al. [56] conducted research on the feasibility of combining photovoltaic and piezoelectric energy harvesting technologies in unmanned air vehicles in 2008. They discovered that an airplane with two piezoelectric cantilever beams on the landing gear and a load of monocrystalline solar cells outperformed the equivalent aircraft without any energy harvesters [57]. The integration of photovoltaic and piezoelectric effects in a dye-sensitized solar cell and a nanogenerator made from ZnO nanowire demonstrated a 6 percent improvement in power output with the introduction of the piezoelectric nanogenerator [57]. In another study, Yu et al. [58] proposed a solar vibration hybrid energy harvester by employing piezoelectric PZT cantilever arrays laminated with Si solar cells for simultaneous harvesting of vibration and solar light. This is particularly important in cases of low-intensity indoor light, which may not be sufficient for powering wireless sensor systems. A power conditioning circuit is used to condition the integrated power output from the hybrid source to reduce the dissipation of power. Multisources can also be a combination of piezoelectric and thermoelectric sources in which heat may be produced when the system is subjected to deformation (e.g. thermoacoustic energy). Lee et al. [59] designed a flexible hybrid cell to concurrently harvest thermal and mechanical energies. ZnO nanowires were used to capture mechanical energy from human body movement and heat. Furthermore, high energy ball milling and spark plasma sintering techniques were used to create n–and p-type thermoelectric materials with good performance utilizing the powders Bi-Te-Se and Bi-Sb-Te, respectively. These n–and p–type thermoelectrics were attached in a series connection electrically while in parallel connection thermally within Al substrates. The Al substrates are good thermal, as well as electrical, conductors. ZnO nanowires were grown on the substrates to act as electrodes. Finally the two devices were stacked together vertically (Figure 7.12). The hybrid cell demonstrated unique advantages of high output current from the thermal generator (≈ 5 μA) and high output voltage from piezoelectric nanogenerator (≈ 3 V). This multisource harvester was proposed for harvesting skin heat and body movement. In a recent report, the piezoelectric output power was increased with thermoelectric assistance. The harvested power increased to approximately 3.4 times and 4 times, respectively, that of the conventional piezoelectric output voltage (8 V) produced by the piezoelectric transducer, with the help of thermoelectric energy (0.1 V and 0.2 V) (PZT). Furthermore, with thermoelectric aid of 0.1 V and 0.2 V, respectively, the gathered power increased by 13.3 percent and 33.3 percent [60].

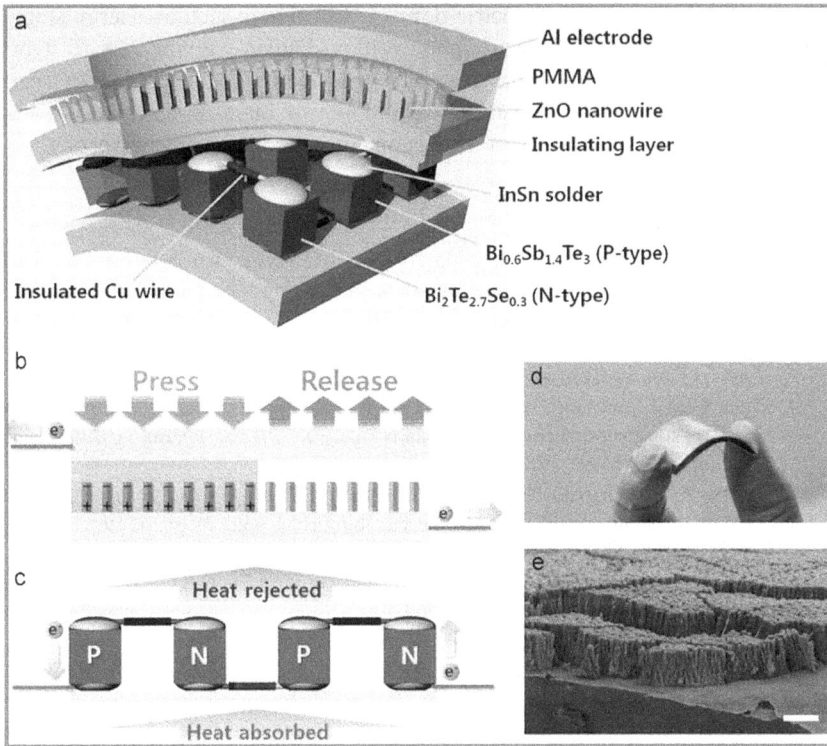

FIGURE 7.12 (a) Illustration showing the construction of hybrid cell of vertically stacked nanogenerator over thermoelectric generator; working principle of the (b) nanogenerator showing polarity induced by applied strain and (c) electron flow diagram of thermoelectric generator; (d) real photograph of flexible hybrid cell consisting of nanogenerator and thermoelectric generator; (e) SEM micrographs of as grown ZnO wires on Al substrate.

Source: [59].

7.8 SUMMARY

Dynamic applications in numerous domains have been made possible by the ongoing development of novel piezoelectric materials with enhanced capabilities (mechanical, electromechanical, thermal, and biocompatibility). These cutting-edge application-based gadgets generate energy from several sources, including fluid flow, the human body, animals, infrastructure, and automobiles. Scavenging energy from fluid flow (water and air) have become crucial in remote areas to power small electronic devices where the availability of other conventional power sources is difficult. Further, wearable and implantable energy-harvesting devices play a very important role in powering small and portable electronics as well medical implants. Above all, multifunctional and multisource energy harvesting technologies integrated piezoelectric domain with others such as energy storage and electromagnetic/thermoelectric/photovoltaic domains, giving rise to hybrid harvesters. However, many of the examples provided in this chapter are currently still in the research stage. A substantial amount of ongoing research on these dimensionalities will lead to major change in cutting-edge energy harvesting.

REFERENCES

1. Federspiel, C.C. and Chen, J., 2003. *Air-powered Sensor* (Vol. 1, pp. 22–25). IEEE.
2. Priya, S., 2005. Modeling of electric energy harvesting using piezoelectric windmill. *Applied Physics Letters*, *87*(18), p. 184101.
3. Myers, R., Vickers, M., Kim, H. and Priya, S., 2007. Small scale windmill. *Applied Physics Letters*, *90*(5), p. 054106.
4. Tien, C.M.T. and Goo, N.S., 2010. Use of a piezo-composite generating element for harvesting wind energy in an urban region. *Aircraft Engineering and Aerospace Technology 82*(6) pp. 376–381.
5. Bressers, S., Avirovik, D., Lallart, M., Inman, D.J. and Priya, S., 2011. Contactless wind turbine utilizing piezoelectric bimorphs with magnetic actuation. In, *Structural Dynamics, Vol. 3* (pp. 233–243). Springer, New York.
6. Yang, Y., Shen, Q., Jin, J., Wang, Y., Qian, W. and Yuan, D., 2014. Rotational piezoelectric wind energy harvesting using impact-induced resonance. *Applied Physics Letters*, *105*(5), p. 053901.
7. Kan, J., Fan, C., Wang, S., Zhang, Z., Wen, J. and Huang, L., 2016. Study on a piezo-windmill for energy harvesting. *Renewable Energy*, *97*, pp. 210–217.
8. Naudascher, E. and Rockwell, D., 1980. Oscillator-Model Approach to the Identification and Assessment of flow-Induced Vibrations in a System. *Journal of Hydraulic Research*, *18*(1), pp. 59–82.
9. Pankanin, G.L., 2005. The vortex flowmeter: various methods of investigating phenomena. *Measurement Science and Technology*, *16*(3), p. R1.
10. Li, S. and Lipson, H., 2009, January. Vertical-stalk flapping-leaf generator for wind energy harvesting. In, *Smart Materials, Adaptive Structures and Intelligent Systems* (Vol. 48975, pp. 611–619).
11. Tan, Y.K. and Panda, S.K., 2007, November. A novel piezoelectric based wind energy harvester for low-power autonomous wind speed sensor. In, *IECON 2007-33rd Annual Conference of the IEEE Industrial Electronics Society* (pp. 2175–2180). IEEE.
12. Li, S., Yuan, J. and Lipson, H., 2011. Ambient Wind Energy Harvesting Using Cross-flow Fluttering. *Journal of Applied Physics 109*(2) p. 026104.
13. Wang, W., He, X., Wang, X., Wang, M. and Xue, K., 2018. A bioinspired structure modification of piezoelectric wind energy harvester based on the prototype of leaf veins. *Sensors and Actuators A: Physical*, *279*, pp. 467–473.
14. Wang, K., Xia, W., Lin, T., Wu, J. and Hu, S., 2021. Low-speed flutter of artificial stalk-leaf and its application in wind energy harvesting. *Smart Materials and Structures*, *30*(12), p. 125002.
15. Laser, D.J., and J.G. Santiago. *J. Micromech. Microeng.* 2004. R35–R64.
16. Song, R., Shan, X., Lv, F., Li, J. and Xie, T., 2015. A novel piezoelectric energy harvester using the macro fiber composite cantilever with a bicylinder in water. *Applied Sciences*, *5*(4), pp. 1942–1954.

17. Song, R., Shan, X., Lv, F. and Xie, T., 2015. A study of vortex-induced energy harvesting from water using PZT piezoelectric cantilever with cylindrical extension. *Ceramics International*, *41*, pp. S768-S773.
18. Shan, X., Song, R., Fan, M. and Xie, T., 2016. Energy-harvesting performances of two tandem piezoelectric energy harvesters with cylinders in water. *Applied Sciences*, *6*(8), p. 230.
19. Xie, X.D., Wang, Q. and Wu, N., 2014. Potential of a piezoelectric energy harvester from sea waves. *Journal of Sound and Vibration*, *333*(5), pp. 1421–1429.
20. Taylor, G.W., Burns, J.R., Kammann, S.A., Powers, W.B. and Welsh, T.R., 2001. The energy harvesting eel: A small subsurface ocean/river power generator. *IEEE Journal of Oceanic Engineering*, *26*(4), pp. 539–547.
21. Techet, A.H., Allen, J.J. and Smits, A.J., 2002, May. Piezoelectric eels for energy harvesting in the ocean. In, *The Twelfth International Offshore and Polar Engineering Conference*. OnePetro.
22. Starner, T., 1996. Human-powered wearable computing. *IBM Systems Journal*, *35*(3.4), pp. 618–629.
23. Fan, K., Yu, B., Zhu, Y., Liu, Z. and Wang, L., 2017. Scavenging energy from the motion of human lower limbs via a piezoelectric energy harvester. *International Journal of Modern Physics B*, *31*(7), p. 1741011.
24. Kymissis, J., Kendall, C., Paradiso, J. and Gershenfeld, N., 1998, October. Parasitic power harvesting in shoes. In, *Digest of Papers. Second International Symposium on Wearable Computers* (Cat. No. 98EX215) (pp. 132–139). IEEE.
25. Xia, H. and Shull, P.B., 2018, June. Preliminary testing of an angled sliding shoe for potential human energy harvesting applications. In, *2018 2nd International Conference on Robotics and Automation Sciences (ICRAS)* (pp. 1–4). IEEE.
26. Cha, Y. and Seo, J., 2018. Energy harvesting from a piezoelectric slipper during walking. *Journal of Intelligent Material Systems and Structures*, *29*(7), pp. 1456–1463.
27. Katsumura, H., Konishi, T., Okumura, H., Fukui, T., Katsu, M., Terada, T., Umegaki, T. and Kanno, I., 2018, July. Development of piezoelectric vibration energy harvesters for battery-less smart shoes. In, *Journal of Physics: Conference Series* (Vol. 1052, No. 1, p. 012060). IOP Publishing.
28. Fu, Y.F., Yi, F.L., Liu, J.R., Li, Y.Q., Wang, Z.Y., Yang, G., Huang, P., Hu, N. and Fu, S.Y., 2020. Super soft but strong E-Skin based on carbon fiber/carbon black/silicone composite: Truly mimicking tactile sensing and mechanical behavior of human skin. *Composites Science and Technology*, *186*, p. 107910.
29. Mokhtari, F., Spinks, G.M., Fay, C., Cheng, Z., Raad, R., Xi, J. and Foroughi, J., 2020. Wearable electronic textiles from nanostructured piezoelectric fibers. *Advanced Materials Technologies*, *5*(4), p. 1900900.
30. Dagdeviren, C., Yang, B.D., Su, Y., Tran, P.L., Joe, P., Anderson, E., Xia, J., Doraiswamy, V., Dehdashti, B., Feng, X., Lu, B., Poston, R., Khalpey Z., Ghaffari, R., Huang, Y., Slepian, M. J. and Rogers, J. A., 2014. Conformal piezoelectric energy harvesting and storage from motions of the heart, lung, and diaphragm. *Proceedings of the National Academy of Sciences*, *111*(5), pp. 1927–1932.
31. J.A. Potkay, K. Brooks, 2008. An arterial cuff energy scavenger for implanted microsystems. The Second International Conference on Bioinformatics and Biomedical Engineering, May, Shanghai, China, pp. 1580–1583
32. Zhang, H., Zhang, X.S., Cheng, X., Liu, Y., Han, M., Xue, X., Wang, S., Yang, F., Smitha, A.S., Zhang, H. and Xu, Z., 2015. A flexible and implantable piezoelectric

generator harvesting energy from the pulsation of ascending aorta: in vitro and in vivo studies. *Nano Energy, 12*, pp. 296–304.

33. Chen, P., Wu, P., Wan, X., Wang, Q., Xu, C., Yang, M., Feng, J., Hu, B. and Luo, Z., 2021. Ultrasound-driven electrical stimulation of peripheral nerves based on implantable piezoelectric thin film nanogenerators. *Nano Energy, 86*, p. 106123.

34. Shafer, M.W. and Morgan, E., 2014, September. Energy harvesting for marine-wildlife monitoring. In, *Smart Materials, Adaptive Structures and Intelligent Systems* (Vol. 46155, p. V002T07A017). American Society of Mechanical Engineers.

35. Qian, F., Liu, M., Huang, J., Zhang, J., Jung, H., Deng, Z.D., Hajj, M.R. and Zuo, L., 2022. Bio-inspired bistable piezoelectric energy harvester for powering animal telemetry tags: Conceptual design and preliminary experimental validation. *Renewable Energy, 187*, pp. 34–43.

36. Reissman, T., MacCurdy, R.B. and Garcia, E., 2008, January. Experimental study of the mechanics of motion of flapping insect flight under weight loading. In, *Smart Materials, Adaptive Structures and Intelligent Systems* (Vol. 43321, pp. 699–709).

37. Reissman, T. and Garcia, E., 2008. Cyborg MAVs using power harvesting and behavioral control schemes. In, *Advances in Science and Technology* (Vol. 58, pp. 159–164). Trans Tech Publications.

38. Shafer, M.W., MacCurdy, R., Shipley, J.R., Winkler, D., Guglielmo, C.G. and Garcia, E., 2015. The case for energy harvesting on wildlife in flight. *Smart Materials and Structures, 24*(2), p. 025031.

39. Li, X. and Strezov, V., 2014. Modelling piezoelectric energy harvesting potential in an educational building. *Energy Conversion and Management, 85*, pp. 435–442.

40. Hwang, S.J., Jung, H.J., Kim, J.H., Ahn, J.H., Song, D., Song, Y., Lee, H.L., Moon, S.P., Park, H. and Sung, T.H., 2015. Designing and manufacturing a piezoelectric tile for harvesting energy from footsteps. *Current Applied Physics, 15*(6), pp. 669–674.

41. Moro, L. and Benasciutti, D., 2010. Harvested power and sensitivity analysis of vibrating shoe-mounted piezoelectric cantilevers. *Smart Materials and Structures, 19*(11), p. 115011.

42. Xie, X.D., Wu, N., Yuen, K.V. and Wang, Q., 2013. Energy harvesting from high-rise buildings by a piezoelectric coupled cantilever with a proof mass. *International Journal of Engineering Science, 72*, pp. 98–106.

43. Xie, X.D., Wang, Q. and Wang, S.J., 2015. Energy harvesting from high-rise buildings by a piezoelectric harvester device. *Energy, 93*, pp. 1345–1352.

44. Jiang, X., Li, Y., Li, J., Wang, J. and Yao, J., 2014. Piezoelectric energy harvesting from traffic-induced pavement vibrations. *Journal of Renewable and Sustainable Energy, 6*(4), p. 043110.

45. Xu, X., Cao, D., Yang, H. and He, M., 2018. Application of piezoelectric transducer in energy harvesting in pavement. *International Journal of Pavement Research and Technology, 11*(4), pp. 388–395.

46. Zhao, X., Xiang, H. and Shi, Z., 2020. Piezoelectric energy harvesting from vehicles induced bending deformation in pavements considering the arrangement of harvesters. *Applied Mathematical Modelling, 77*, pp. 327–340.

47. Zhang, Z., Xiang, H., Shi, Z. and Zhan, J., 2018. Experimental investigation on piezoelectric energy harvesting from vehicle-bridge coupling vibration. *Energy Conversion and Management, 163*, pp. 169–179.

48. Múčka, P., 2016. Energy-harvesting potential of automobile suspension. *Vehicle System Dynamics*, *54*(12), pp. 1651–1670.

49. Khameneifar, F. and Arzanpour, S., 2008, January. Energy harvesting from pneumatic tires using piezoelectric transducers. In, *Smart Materials, Adaptive Structures and Intelligent Systems* (Vol. 43314, pp. 331–337).

50. Singh, K.B., Bedekar, V., Taheri, S. and Priya, S., 2012. Piezoelectric vibration energy harvesting system with an adaptive frequency tuning mechanism for intelligent tires. *Mechatronics*, *22*(7), pp. 970–988.

51. Maurya, D., Kumar, P., Khaleghian, S., Sriramdas, R., Kang, M.G., Kishore, R.A., Kumar, V., Song, H.C., Park, J.M.J., Taheri, S. and Priya, S., 2018. Energy harvesting and strain sensing in smart tire for next generation autonomous vehicles. *Applied Energy*, *232*, pp. 312–322.

52. Anton, S.R., Erturk, A. and Inman, D.J., 2010. Multifunctional self-charging structures using piezoceramics and thin-film batteries. *Smart Materials and Structures*, *19*(11), p. 115021.

53. Anton, S.R., Erturk, A. and Inman, D.J., 2012. Multifunctional unmanned aerial vehicle wing spar for low-power generation and storage. *Journal of Aircraft*, *49*(1), pp. 292–301.

54. Wang, Y. and Inman, D.J., 2013. Simultaneous energy harvesting and gust alleviation for a multifunctional composite wing spar using reduced energy control via piezoceramics. *Journal of Composite Materials*, *47*(1), pp. 125–146.

55. Sugino, C. and Erturk, A., 2018. Analysis of multifunctional piezoelectric metastructures for low-frequency bandgap formation and energy harvesting. *Journal of Physics D: Applied Physics*, *51*(21), p. 215103.

56. Magoteaux, K.C., Sanders, B. and Sodano, H.A., 2008, April. Investigation of an energy harvesting small unmanned air vehicle. In, *Active and Passive Smart Structures and Integrated Systems 2008* (Vol. 6928, pp. 610–620). SPIE.

57. Xu, C. and Wang, Z.L., 2011. Compact hybrid cell based on a convoluted nanowire structure for harvesting solar and mechanical energy. *Advanced Materials*, *23*(7), pp. 873–877.

58. Yu, H., Yue, Q., Zhou, J. and Wang, W., 2014. A hybrid indoor ambient light and vibration energy harvester for wireless sensor nodes. *Sensors*, *14*(5), pp. 8740–8755.

59. Lee, S., Bae, S.H., Lin, L., Ahn, S., Park, C., Kim, S.W., Cha, S.N., Park, Y.J. and Wang, Z.L., 2013. Flexible hybrid cell for simultaneously harvesting thermal and mechanical energies. *Nano Energy*, *2*(5), pp. 817–825.

60. Chen, Z., Xia, Y., Shi, G., Xia, H., Wang, X., Qian, L. and Ye, Y., 2021. Enhanced piezoelectric energy harvesting power with thermoelectric energy assistance. *Journal of Intelligent Material Systems and Structures*, *32*(18–19), pp. 2260–2272.

8 Summary and Future Scope

8.1 SUMMARY

There has been a profound influence of material technologies on the lifestyle of human civilization for ages. With time, we have evolved from the Stone Age to the age of smart materials depending on the necessity of new inventions and the quest to lead a comfortable life. Piezoelectrics are a group of smart materials that can transform mechanical stress into electric voltage and vice versa. Such an intelligent response of the piezoelectric materials mimicking bionic systems of humans and animals with a unique combination of sensory network (e.g. skin sensing thermal gradients; eye sensing optical signals) makes them suitable for diversified applications. Hence, this monograph is devoted to the understanding of various aspects of piezoelectric energy harvesting, starting from its figures of merit to materials and applications. Moreover, the growing demand for green energy harvesting has been the motivation for the development of this monograph to present the current status of piezoelectric research in this direction. The book has been divided systematically into eight chapters, each highlighting different ingredients necessary for piezoelectric energy harvesting.

Chapter 1 provides an introduction to green energy harvesting and its necessity in the present scenario. It describes energy harvesting from various ambient sources, including solar, wind, waves, geothermal and artificial sources such as piezoelectric, thermoelectric, magnetostrictive and magnetoelectric. This is necessary for the development of stand-alone systems, for example, wireless sensor networks in remote locations, wearable and mobile electronics and biomedical devices. Further, the practical challenges faced while extracting energy from ambient resources and future prospectives of artificial sources are discussed.

Chapter 2 illustrates the fundamentals of piezoelectricity, direct and indirect piezoelectric effects, and their usefulness in the electro-ceramic industry. Domain structure and domain wall movement and the corresponding polarization hysteresis loop as a function of an electric field are discussed in detail. Further, the role of different figures of merit (piezoelectric charge constant, voltage constant, electromechanical coupling factor, mechanical quality factor, permittivity, output

 DOI: 10.1201/9781003317289-8

power, etc.) in designing materials specific for different applications are also debated. A roadmap to achieve superior values of figures of merit is described in terms of construction of morphotropic phase boundaries (MPB) and introduction of relaxor behavior. MPB usually refers to the composition-dependent phase boundaries around which two phases are energetically equal but structurally different. The second feature, that is, relaxor behavior arises from the structural disorder within materials having disruption of long range systematic atomic arrangements due to defects/imperfections. Such imperfections lead to perturbation in the structural symmetry that is responsible for functional response of technologically advanced materials.

Chapter 3 presents an extensive review on the different types of piezoelectric bulk and nanostructured materials including ceramics, polymers, polymer/ ceramic composites, thin films, bio-inspired materials suitable for energy harvesting applications. This wide spectrum of piezoelectric materials braces their own importance in terms of figures of merit. For instance, ceramics posses high piezoelectric constants but are highly brittle. These ceramic materials can be blended with piezo-polymers resulting in flexible composites. Furthermore, ceramic materials are classified as polycrystalline, single crystal, textured, nanostructured and thin film materials while piezo-composites were categorized as 0-3, 1-3, 3-3 composites based on their piezoelectricity. Bio-inspired materials are described as an emerging class of piezo-materials that not only reduce biological wastes but also limit toxic e-wastes from electronic industries.

Chapter 4 narrates the importance of various synthesis techniques for tailoring the properties of piezoelectric materials specific to particular applications. Synthesis methodology is very crucial in modulating the particle size, morphology, formation of secondary phases, density, and so forth, of the materials. With this viewpoint, some of the essential synthesis techniques such as powder processing routes, chemical synthesis routes, thin film deposition, chemical solvent deposition, melt growth techniques, solution growth, dielectrophoresis, injection molding, lost foam mold and dice and fill techniques are discussed in detail. Synthesis/ fabrication techniques are chosen basing on the material requirement for cutting edge applications.

Chapter 5 emphasizes several characterization techniques necessary for studying the various properties of piezoelectric materials. Following the synthesis of materials, it is the foremost concern of the researchers to characterize the samples in order to investigate and analyze their behavior. For instance, to confirm the phase, structure, lattice strain, particle size, lattice parameters and other crystallographic details, one needs to perform XRD and Raman spectroscopic analysis. For morphological studies, electron microscopic techniques such as scanning electron microscopy (SEM), transmission electron microscopy (TEM) and atomic force microscopy (AFM) are important. Thermal analysis techniques such as thermo-gravimetric analysis (TGA), differential thermal analysis (DTA), differential scanning calorimetry (DSC) are essential in studying phase transitions, glass transition temperature, crystalline kinetics of glass and polymers, estimate heat capacity, coefficient of thermal expansion, enthalpy and so forth. Apart from the basic characterization techniques, dielectric study, P-E (polarization versus electric field) hysteresis loop, impedance study and measurement of electromechanical properties such as strain versus electric field (S-E) loop, piezoelectric and coupling constants are of utmost importance. With this motive, the chapter summarizes various important characterization methods necessary for piezoelectric study.

Chapter 6 describes the importance of piezoelectric structures in energy harvesting and fundamental mechanism responsible for their working. The complex piezoelectric energy harvesting technology involves synchronization among different segments by proper choice of key parameters. Hence, the structural configuration and operation modes play a very important role in building an appropriate piezoelectric energy harvester. Considering such factors, important configurations such as unimorph/bimorph cantilever, cymbal structure, circular diaphragms, and stacked harvesters are discussed in detail. Further, the operation of these configurations in 31–and 33–vibration modes are reviewed.

Chapter 7 bridges the various piezoelectric structural configurations and vibrational modes to applications and devices. These structures can be employed to harness energy from ambient sources such as fluid flow, industrial machinery, buildings/bridges, movement of human body parts, including eyelid, skin, limbs and heartbeat; animals (flapping of wings; limb/neck movement) vehicles and so forth. Multifunctional and multi-sources that integrate piezoelectric energy harvesting with energy storage and electromagnetic/thermoelectric/photovoltaic domains are also included in the chapter.

Chapter 8, that is, the present chapter, provides a brief summary of all the chapters and also discusses the challenges and future scope of piezoelectric energy harvesters.

8.2 CHALLENGES AND FUTURE SCOPE

With the maturation of technology, the hunger for sustainable energy sources is increasing day by day. In view of this, piezoelectric devices seem to be promising for the conversion of parasitic energy into useful energy. However, in spite of encouraging results, only a few piezoelectric products are prevalent in the market. One of the factors might be the toxicity issues related to high-performance, lead-based piezolelectrics and lack of availability of lead-free systems with properties at par with Pb-family. It is usually noticed that the Pb-based group offers much superior piezoelectric properties for application purposes. However, the release of hazardous waste such as PbO during the synthesis of these materials keeps their use limited. Although new materials in the form of lead-free alternatives have emerged in recent days, their properties are yet to reach the standards needed for commercialization and production on a large scale. So, it is a real challenge for material scientists and industries to develop environmentally benign materials with optimum features. Apart from that, biocompatibility and biodegradability of the piezoelectric materials still remains a concern. The in vivo efficacy as well as prolonged in vivo stability of piezoceramics – including $BaTiO_3$, ZnO, $BiFeO_3$; piezopolymers such as PVDF and poly(lactic–co-glycolic) acid (PLGA) – and even composites are yet to be established. For instance, in drug-delivery systems, magnetically driven piezoelectric nanocarriers are found to be effective in vitro. But drug loading efficiency and site-specific targeting need a lot of improvement when examined in vivo. Therefore, it is of utmost importance to perform clinical translation to examine the biocompatibility, biodegradability, and tissue accumulation of the piezo-materials when used in biological applications. Besides, a big concern of the piezoelectric energy harvesters is the effective design, which can deliver maximum power at a low cost. Currently, serious attention is being diverted in this direction to constructing suitable cost-effective piezoelectric architectures. Textured piezo-materials may be a good choice for future PEH devices. This category of materials demonstrates a combination of high d_{33} with low dielectric constant, which is one of the potential ways of enhancing power output. Furthermore, construction of hybrid devices for simultaneous scavenging of multiple-type energies can be one of the beneficial means to enhancing the output performance of piezoelectric energy harvesters. Hence, hybrid cells can be designed by combining piezoelectric technology with photovoltaic, thermoelectric, biochemical, and biomechanical energies. For example, innovative approaches can be adopted to harvest energy from mechanical movements along with thermal energy from body heat can be expected to improve the generation of energy. On the other hand, research on integrating vibration energy harvesting with storage via electrochemical capacitors on the same device is still in its infancy and, if developed, can drive technology to the next level. Thus, it can be anticipated that with a proper blueprint, PEHs can open up new avenues for sustainable green energy harvesting to power most wireless sensors and other electronics with improved performance.

Index

For Product Safety Concerns and Information please contact our EU
representative GPSR@taylorandfrancis.com
Taylor & Francis Verlag GmbH, Kaufingerstraße 24, 80331 München, Germany